Chemical
Information
MINING

Facilitating Literature-Based Discovery

Chemical Information
MINING

Facilitating Literature-Based Discovery

Edited by
DEBRA L. BANVILLE

CRC Press
Taylor & Francis Group
Boca Raton London New York

CRC Press is an imprint of the
Taylor & Francis Group, an **informa** business

CRC Press
Taylor & Francis Group
6000 Broken Sound Parkway NW, Suite 300
Boca Raton, FL 33487-2742

First issued in paperback 2019

© 2009 by Taylor & Francis Group, LLC
CRC Press is an imprint of Taylor & Francis Group, an Informa business

No claim to original U.S. Government works

ISBN-13: 978-1-4200-7649-3 (hbk)
ISBN-13: 978-0-367-38620-7 (pbk)

Library of Congress Cataloging-in-Publication Data

Chemical information mining : facilitating literature-based discovery / Debra L. Banville, editor.
        p. cm.
    Includes bibliographical references and index.
    ISBN 978-1-4200-7649-3 (alk. paper)
    1. Chemical literature--Research. 2. Data mining. 3. Information storage and retrieval systems--Chemistry. I. Banville, Debra L.

QD8.5.C475 2009
025.06'54--dc22   2008030749

Visit the Taylor & Francis Web site at
http://www.taylorandfrancis.com

and the CRC Press Web site at
http://www.crcpress.com

Contents

Part I Introduction to Information Mining for the Life Sciences

Part II Chemical Semantics

Part III   Trends in Chemical Information Mining

Part IV   Involving the Researchers and Closing the Loop

Preface

Researchers in scientific endeavors such as the life sciences frequently face significant concerns when looking for the most relevant or "right" information from the literature. For example, what happens when we find too much information on a subject, cannot find any information, or cannot access the full text documents when interesting citations are found? These are concerns that most life science researchers face every day but rarely acknowledge. The magnitude of the problem is most commonly expressed as a growing interest in text extraction capabilities and our use of web search engines such as Google, and PubMed or PubChem to provide easy awareness of scientific life science information.

This book is about acknowledging concerns of information extraction, highlighting solutions available today, and underscoring the value these solutions bring to both academic and commercial scientists alike. A special focus is on chemical information extraction due to its importance in so many life science areas and to fill a gap in the literature that still exists at the time this book is being written. Chemical entity extraction is meant to complement the extensive literature on biological entity extraction. The ultimate goal, as described in this book, is to build relationships between chemical and biological entities—relationships that are at the heart of life science research.

The intent of this book is also holistic: to look at both the technological details, in this case the development of chemical structure extraction capabilities, and to provide a possible road map for how researchers can best think about these technologies in their daily work. On one hand, a road map is meant to underscore to developers that the ability to provide a great chemical text extraction capability is most valuable when the scientists needing this capability are factored into the process. On the other hand, we want to underscore to researchers that the capabilities of chemical text mining present new opportunities in how researchers think about and manage their information, and this requires openness to new techniques and capabilities. Ideally, those developing these new capabilities and the researchers needing those capabilities can collaborate on shaping the future of scientific information and knowledge management. This book is written with this vision in mind.

Acknowledgments

To all my colleagues at AstraZeneca who have supported me throughout the years I give thanks, with special thanks to Jim Rosamond and Jim Damewood, who I have worked very closely with in the area of chemical information mining. To the contributors of this book, thank you for your invaluable contributions! I could not have done this book without your generous gift of time and effort. We have something to be very proud of. A thanks also goes to the publishers for believing in this book. Finally, to my husband Don and daughter Janice, thank you for allowing me to take valuable personal time away from you to write and edit this book.

The Editor

Debra L. Banville is currently a scientific information analyst at AstraZeneca Pharmaceuticals within the R&D areas of drug discovery. Debra's focus is primarily on improving the management of both scientific information and knowledge about that information. In recognition of her work, Debra and her team were the recipients of an internal award for innovation. Debra has been an invited speaker at several conferences including most recently the Infonortics conference in Spain (October 2007) and the PharmaBioMed Conference in Portugal (2006), and an invited author for *Drug Discovery Today* (January 2006). Prior to her work in information science, Debra ran an active research program in the areas of drug binding to biological targets using multidimensional nuclear magnetic resonance techniques. Debra received her B.S. from Brandeis University, her Ph.D. from Emory University, and her post-doctorate at the University of California at San Francisco.

Contributors

Debra L. Banville
AstraZeneca Pharmaceuticals
Wilmington, Delaware
debra.banville@astrazeneca.com

Colin Batchelor
Informatics Department
Royal Society of Chemistry
Cambridge, United Kingdom
batchelorc@rsc.org

Roger Beckman
Chemistry Library
Indiana University
Bloomington, Indiana
beckmanr@indiana.edu

Martin Hofmann-Apitius
Fraunhofer Institute for
  Algorithms and Scientific
  Computing (SCAI)
Sankt Augustin, Germany
martin.hofmann-apitius@scai.fhg.de

A. Peter Johnson
School of Chemistry
University of Leeds
Leeds, United Kingdom
p.johnson@leeds.ac.uk

Richard Kidd
Informatics Department
Royal Society of Chemistry
Cambridge, United Kingdom
kiddr@rsc.org

Corinna Kolářik
Fraunhofer Institute for
  Algorithms and Scientific
  Computing (SCAI)
Sankt Augustin, Germany
corinna.kolarik@scai.fraunhofer.de

Bedřich Košata
Laboratory of Informatics and
  Chemistry
Institute of Chemical Technology
Prague, Czech Republic
bedrich.kosata@vscht.cz

Miloslav Nic
Laboratory of Informatics and
  Chemistry
Institute of Chemical Technology
Prague, Czech Republic
miloslav.nic@vscht.cz

Anikó T. Valkó
Keymodule Ltd.
Leeds, United Kingdom
aniko.valko@keymodule.co.uk

David J. Wild
School of Informatics
Indiana University
Bloomington, Indiana
djwild@indiana.edu

Antony J. Williams
ChemZoo Inc. and
  ChemConnector Inc.
Wake Forest, North Carolina
antony.williams@chemspider.com

Andrey Yerin
Advanced Chemistry
  Development Inc.
Moscow, Russian Federation
yerin@acdlabs.ru

Part I

Introduction to Information Mining for the Life Sciences

1 Illustrating the Power of Information in Life Science Research

Debra L. Banville

CONTENTS

INTRODUCTION

The ironic proverbial saying that "a month in the lab can save you an hour in the library" is proving itself repeatedly and at a huge cost to both academic and commercial institutions alike. Missed information in the literature costs time, money, and quality. Both the quality of decisions made and the quality of subsequent research output is compromised when the available information is not realized. In monetary terms, incorrect decisions along the drug pipeline lifecycle in the pharmaceutical area can cost millions to billions of dollars (Adams and Brantner 2006; Banik and Westgren 2004; DiMasi 2002; DiMasi et al. 2003; Gaughan 2006; Leavitt 2003; Myers and Baker 2001).

Substantial costs have been experienced in academia as well and seen as missed funding opportunities due to a combination of access limitations to the information together with the inability to find and process the available information (Wilbanks and Boyle 2006). Access limitations are worse in academia than in industry. Lowering the barriers to access limitations has been the goal of individuals such as Paul Ginsparg, who in 1991 developed arXiv (Ginsparg 1991), the first free scientific online archive of non–peer reviewed physics articles that continues today (Ginsparg et al. 2004). Many groups have formed to increase the accessibility of academic information such as SPARC (Scholarly Publishing and Academic Resources Coalition), the Science Commons group (www.sciencecommons.org), and the World Wide Web Consortium (w3c.org).

The value of information mining the literature for knowledge has been illustrated repeatedly. In 1986, Donald R. Swanson, an information scientist, mathematician,

3

and professor emeritus at the University of Chicago, demonstrated the technique by using the literature to find a possible treatment for Raynaud's syndrome (Swanson 1986, 1987, 1988). Swanson went on to clinically prove the hypothesis suggested by the literature to use fish oils as a treatment for Raynaud's. This work set off a string of papers in an area coined as "literature-based discovery" or "literature-related discovery" (Smalheiser and Swanson 1994, 1996a, b, 1998; Swanson 1990, 1991; Swanson and Smalheiser 1999; Gordon and Lindsay 1996; Kostoff 2007; Weeber et al. 2001).

In fact, literature-based discovery and text mining of the literature are part of the same thing; they are both about extracting information from text to discover something new, novel, or not already known. Text mining and literature-based discovery go beyond the simple analysis of text. Ideally they led to the recognition of interesting patterns not explicitly stated. The most recent and prominent example of this, at the writing of this book, is a January 2008 article by a group from Peking University (Li et al. 2008). These researchers asked the question, Is there a common molecular pathway in addiction? They first identified ~1,000 relevant articles on the subject and manually extracted 2,343 items of evidence. They kept only well-established evidence and extensively annotated and then stored this evidence in a searchable database for further analysis. Based on their meticulous extraction and analysis, they identified five molecular pathways common to four different types of addictive drugs. This included discovering two new pathways and clues to the irreversible features of addiction. They did this without conducting a single experiment.

A rigorous description of literature-based discovery was published by Kostoff in an earlier paper (Kostoff 2007) and followed later by a series of eight papers that detailed the techniques used and demonstrated these techniques for a variety of life science areas including cataracts, Raynaud's, Parkinson's, and multiple sclerosis, and water purification (Kostoff 2008a,b; Kostoff et al. 2008a–f).

Other opportunities for knowledge discovery from the literature includes the area of *drug repurposing*, the development of novel uses for existing drugs. Most drug repurposing (also known as drug *reprofiling* or *repositioning*) discoveries were the result of researchers connecting key information to generate a valid hypothesis that could be tested in the clinic (Wilkinson 2002; Lipinski 2006; Oprea and Tropsha 2006; Ashburn and Thor 2004). Repurposed drugs frequently have the advantage of having been previously tested in the clinic for safety and are simply being reapplied to a novel area. This is not a new concept: in 2004, 84% of the 50 top-selling drugs had additional indications approved since their launch in the United States (Kregor 2007). For example, two drugs on the market for Parkinson's disease, Ropinirole/Requip (GSK) and Pramipexole/Mirapex (Boehringer Ingelheim), were later repurposed for restless leg syndrome. Repurposing involves additional expense for phase IV trials to support the new indication, application, and marketing fees, but this is nothing compared to the cost of running phase I, II, and III trials.

The bottom line is that in the hands of creative, experienced researchers, text mining of the literature or literature-based discovery can only serve to increase the opportunities within drug discovery and enhance life science research. Turning the ironic proverbial phrase around to read "an hour in the library saves a month in the lab" would be a more advisable approach.

BARRIERS TO THE AUTOMATION OF THESE DISCOVERIES

Finding relevant information involves finding relevant documents, accessing those documents, and finding relevant information within those documents. Numerous barriers exist along each step of the way (Banville 2006 and references within, 2008). For example, imagine that you are trying to find all the bicyclic compounds known to be selective for a specific target. What are some of the issues you would encounter?

- Too many sources to search
- No structure searching capability available within most of these sources
- Limited accessibility to all the necessary sources due to licensing costs
- Limited rights to download and manage the citations and documents found due to licensing restrictions

Getting a citation from PubMed, for example, does not mean that the scientist has access to the full text document cited or the right to use a computer to mine a large set of full text documents. Controversies over the announcement that researchers supported by the National Institute of Health (NIH) will be required to submit all peer-reviewed articles to the NIH for public access within 12 months of publication has predictably drawn positive reviews from most researchers and negative reviews from most publishers (Morrissey 2008). Even if full text access is available, the logistics of downloading all 100 or 1,000 "must read" full text articles are tedious, to say the least. For example, the group from Peking University engaged many students over two years to read, extract, and annotate information from 1,000 documents relevant to drug addiction (Li et al. 2008 and correspondence with L. Wei).

THERE HAS GOT TO BE A BETTER WAY TO DO THIS — SHIFTING PARADIGMS

A variety of technological advances including the advent of the Semantic Web and social networking are driving a cultural change in how information is found and presented back to the user (e.g., Murray-Rust et al. 1997; Rzepa 1998; Berners-Lee et al. 2001; Luo 2007; Chang 2007; Dong et al. 2007). It is no longer about publishing information in print form; it is about *ePublishing* with the ability for communities of readers to comment on this information. This effectively captures knowledge about information, a new paradigm in information sharing. It is also about automatic linking to related information as a form of knowledge sharing and knowledge building.

Publishers like the Royal Society of Chemistry have initiated *Project Prospect* to enhance articles prior to publication with chemical and biological concepts (see rsc. org and *The Alchemist Newsletter* 2007 for details). Chapter 8 of this volume has a detailed discussion on publishing. The ability to find chemical structural information and its associated data is becoming much easier as the result of these endeavors and their many contributors (such as Rupp et al. 2007; Wilkinson 2002; Corbett et al. 2007; Batchelor and Corbett 2007; Corbett and Murray-Rust 2006; Nic et al. 2002; Murray-Rust and Rzepa 1999; Zimmermann and Hofmann 2007 and references within; Zimmerman et al. 2005; Williams 2005; Williams and Yerrin 1999; Rouse and Beckman 1998; Ibison et al. 1992, 1993a,b; Simon and Johnson 1997 and reference within).

Project Prospect endeavors to use and build acceptance of standards for chemical information by using the International Chemical Identifiers (InChIs) created by the International Union of Pure and Applied Chemistry (IUPAC) as a way to provide a nonproprietary way to make chemical information more machine-readable. To illustrate the potential of this in the simplest way, an InChI for benzene (i.e., **InChI=1/ C6H6/c1-2-4-6-5-3-1/h1-6H**) was pasted into a Google search bar (www.google. com), this resulted in 37 hits in the fall of 2007 and over 1,000 hits 6 months later in the spring of 2008. The top hits were directed at the IUPAC Gold Book as shown in Figure 1.1.

Similar searches on common drugs resulted in many highly relevant hits. In the case of aspirin, shown in Figure 1.1, the links were made to several open-access chemical databases such as The Carcinogenic Potency Project database (http:// potency.berkeley.edu/chempages/ASPIRIN.html), PubChem (http://pubchem.ncbi. nlm.nih.gov/summary/summary.cgi?cid=2244), Drug Bank (http://www.drugbank. ca/cgi-bin/getCard.cgi?CARD=DB00945.txt), and ChemSpider (http://www.chem spider.com/RecordView.aspx?id=2157). While this search does not provide a definitive capability and does not ensure a high degree of accuracy in the results found for these drugs, it does demonstrate the current ability we all have to perform a chemical structure search against a large body of information, the Internet, and retrieve highly relevant results.

Integration of select Internet resources, such as the public chemical databases mentioned above, provides a very practical approach to structure searching the Internet and internal resources (Dong et al. 2007). Chapter 8 elaborates on this concept. As summarized in Chapter 2, another facet of chemical structure mining involves finding information within full text documents that do not traditionally contain identifiers like InChI or SMILE strings. Chapter 5 contains an in-depth discussion of these identifiers.

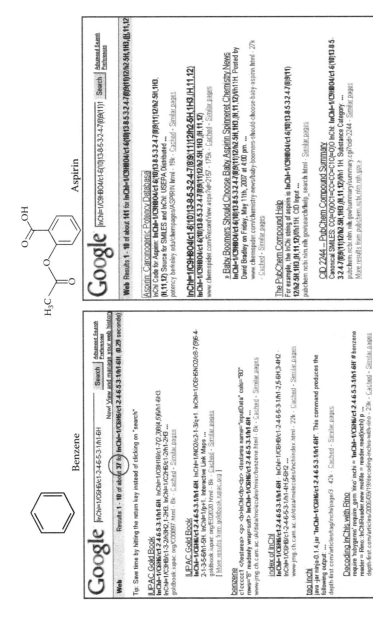

FIGURE 1.1 Google searches on the InChI chemical identifiers for benzene (left) and aspirin (right) demonstrate the potential of structure searching the Web.

REFERENCES

Adams, C. and Brantner, V. 2006. Estimating the cost of new drug development: is it really 802 million dollars? *Health Affairs,* 25(2):420–428.

The Alchemist Newsletter. 2007. Meta data prospecting wins award, grants and awards (September 26). Royal Society of Chemistry's Project Prospect, http://www.rsc.org/Publishing/Journals/ProjectProspect/index.asp.

Ashburn, T.T. and Thor, K.B. 2004. Drug repositioning: identifying and developing new uses for existing drugs. *Nature Reviews Drug Discovery,* 3(8):673–683.

Banik, M. and Westgren, R.E. 2004. A wealth of failures: sensemaking in a pharmaceutical R&D pipeline. *International Journal of Technology Intelligence and Planning,* 1(1):25–38.

Banville, D.L. 2006. Mining chemical structural information from the drug literature. *Drug Discovery Today,* 11(1/2):35–42.

Banville, D.L. 2008. Mining chemical structural information from the literature, in *Pharmaceutical Data Mining: Approaches and Applications for Drug Discovery,* K.V. Balakin (Ed.), Wiley Inc., chap. 20.

Batchelor, C.R. and Corbett, P.T. 2007. Semantic enrichment of journal articles using chimerical named entity recognition. Proceedings of the ACL 2007, demo and poster sessions, Prague, June 2007, pp. 45–48.

Berners-Lee, T., James H., and Ora, L. 2001 (May 17). The semantic web. *Scientific American Magazine.* Available at http://www.sciam.com/article.cfm?id=00048144-10D2-1C70-84A9809EC588EF21&print=true.

Chang, H. 2007. Social networks impact the drugs physicians prescribe according to Stanford Business School research. PharmaLive: business wire (March 16). Available at http://www.gsb.stanford.edu/news/research/mktg_nair_drugs.shtml.

Corbett, P., Batchelor, C., and Teufel, S. 2007. Annotation of chemical named entities. BioNLP 2007: Biological, Translational, and Clinical Language Processing, Prague, June 2007, pp. 57–64.

Corbett, P.T. and Murray-Rust, P. 2006. High-throughput identification of chemistry in life science texts, in *Computational Life Science II,* Lecture Notes in Computer Science series, Vol, 4216. Berlin/Heidelberg: Springer, pp. 107–118.

DiMasi, J. 2002. The value of improving the productivity of the drug development process: faster times and better decisions. *Pharmacoeconomics,* 20(S3):1–10.

DiMasi, J., Hansen, R., and Grabowski, H. 2003. The price of innovation: new estimates of drug development costs. *Journal of Health Economics,* 22(2):151–185.

Dong, X., Gilbert, K.E., Guha, R., Heiland, R., Kim, J., Pierce, M.E., Fox, G.C., and Wild, D.J. 2007. Web service infrastructure for chemoinformatics *Journal of Chemical Information and Modeling,* 47(4):1303–1307.

Gaughan, A. 2006. Bridging the divide: the need for translational informatics. *Future Medicine,* 7(1):117–122.

Ginsparg, P. 1991. Developed arXiv, the first free scientific online archive of non-peer reviewed physics articles. Available at http://www.arXiv.org.

Ginsparg, P., Houle, P., Joachims, T., and Sul, J.-H. 2004. Mapping subsets of scholarly information. *Proceeding of the National Academy of Sciences of the United States of America,* 101:5236–5240.

Gordon, M.D. and Lindsay, R.K. 1996. Toward discovery support systems: A replication, re-examination, and extension of Swanson's work on literature-based discovery of a connection between Raynaud's disease and fish-oil. *Journal of the American Society for Information Science,* 47(2):116–128.

Ibison, P., Jacquot, M., Kam, F., Neville, A.G., Simpson, R.W., Tonnelier, C., Venczel, T., and Johnson, A.P. 1993a. Chemical literature data extraction: the CliDE Project. *Journal of Chemical Information and Computer Science,* 33(3):338–344.

Ibison, P., Jacquot, M., Kam, F., Neville, A.G., Simpson, R.W., Tonnelier, C., Venczel, T., and Johnson, A.P. 1993b. Chemical literature data extraction: the CLiDE Project. *Journal of Chemical Information and Computer Science*, 33(3):338–344.

Ibison, P., Kam, F., Simpson, R.W., Tonnelier, C., Venczel, T., and Johnson, A.P. 1992. Chemical structure recognition and generic text interpretation in the CLiDE Project. Proceedings on Online Information 92, London.

Kostoff, R.N. 2007. Validating discovery in literature-based discovery. *Journal of Biomedical Informatics* 40(4):448–450.

Kostoff, R.N. 2008a. Literature-related discovery (LRD): potential treatments for cataracts. *Technological Forecasting and Social Change,* 75(2):215–225.

Kostoff, R.N. 2008b. Literature-related discovery (LRD): potential treatments for Parkinson's disease. *Technological Forecasting and Social Change,* 75(2):226–238.

Kostoff, R.N., Block, J.A., Solka, J.L., Briggs, M.B., Rushenberg, R.L., Stump, J.A., Johnson, D., Lyons, T.J. and Wyatt, J.R. 2008a. Lessons learned, and future research directions. *Technological Forecasting and Social Change,* 75(2):276–299.

Kostoff, R.N., Block, J.A., Stump, J.A., and Johnson, D. 2008b. Literature-related discovery (LRD): potential treatments for Raynaud's phenomenon. *Technological Forecasting and Social Change,* 75(2):203–214.

Kostoff, R.N., Briggs, M.B., and Lyons, T.J. 2008c. Literature-related discovery (LRD): potential treatments for multiple sclerosis. *Technological Forecasting and Social Change,* 75(2):239–255.

Kostoff, R.N., Briggs, M.B., Solka, J.L., and Rushenberg, R.L. 2008d. Literature-related discovery (LRD): introduction and background. *Technological Forecasting and Social Change,* 75(2):165–185.

Kostoff, R.N., Briggs, M.B., Solka, J.L., and Rushenberg, R.L. 2008e. Literature-related discovery (LRD): methodology. *Technological Forecasting and Social Change,* 75(2):186–202.

Kostoff, R.N., Solka, J.L., Rushenberg, R.L., and Wyatt, J.R. 2008f. Literature-related discovery (LRD): water purification. *Technological Forecasting and Social Change,* 75(2):256–275.

Kregor, P.S. 2007. Drug repurposing. The MSI Consultantancy. Available at http://www.msi.co.uk/article-read.php?DL_ID=37&from=articles.

Leavitt, P.M. 2003. The role of knowledge management in new drug development. *American Productivity & Quality Center*, ProvidersEdge.com:1–7.

Li, C., Mao, X., and Wei, L. 2008. Genes and (common) pathways underlying drug addiction. *PLoS Computational Biology,* 4(1):e2. doi:10.1371/journal.pcbi.0040002 Available at http://compbiol.plosjournals.org/archive/1553-7358/4/1/pdf/10.1371_journal.pcbi.0040002-S.pdf.

Lipinski, C.A. 2006. Why repurposing works and how to pick a winning drug while avoiding failures. *CBI 2nd Annual Conference on Drug Repurposing,* January 30, 2006, Philadelphia, Pennsylvania. Available at http://www.meliordiscovery.com/press/Lipinski_Melior_CBI_Repurposing2006.pdf.

Luo, J.S. 2007. Social networking: now professionally ready. *Primary Psychiatry,* 14(2):21–24. Available at http://www.primarypsychiatry.com/aspx/articledetail.aspx?articleid=975.

Morrissey, S. 2008. Mandatory open access. News of the Week. *Chemical & Engineering News,* 86(3):10.

Murray-Rust, P. and Rzepa, H.S. 1999. Chemical markup, XML and the Worldwide Web. 1. Basic principles. *Journal of Chemical Information and Computer Science,* 39:928–942.

Murray-Rust, P., Rzepa, H.S., and Whitaker, B.J. 1997. The World Wide Web as a chemical information tool. *Chemical Society Reviews,* 26:1–10.

Myers, S. and Baker, A. 2001. Drug discovery—an operating model for a new era. *Nature Biotechnology,* 19:727–730.

Nic, M., Jirat, J., and Kosata, B. 2002. *Compendium of Chemical Terminology* (also known as the *IUPAC Gold Book*). Prague: ICT Press. Available at http://goldbook.iupac.org/index.html.

Oprea, T.I. and Tropsha, A. 2006. Target, chemical and bioactivity databases—integration is key. *Drug Discovery Today: Technologie*, 3(4):357–365.

Rouse, K. and Beckman, R. 1998. Beilstein's CrossFire: a milestone in chemical information and interlibrary cooperation in academia, in *The Beilstein System: System: Database and Software*, S. Heller (Ed.), Washington, DC: American Chemical Society, pp. 133–148.

Rupp, C.J., Copestake, A., Corbett, P., and Waldron B. 2007. Integrating general-purpose and domain-specific components in the analysis of scientific text. Proceedings of the UK e-Science Programme All Hands Meeting (AHM2007), Nottingham, UK. Available athttp://www.allhands.org.uk/2007/proceedings/papers/860.pdf.

Rzepa, H.S. 1998. A history of hyperactive chemistry on the web: From text and images to objects, models and molecular components. *Chimia International Journal for Chemistry*, 52(11):653–657.

Simon, A. and Johnson, A.P. 1997. Recent advances in the CliDE project: logical layout analysis of chemical documents. *Journal of Chemical Information and Computer Science*, 37:109–116.

Smalheiser, N.R. and Swanson, D.R. 1994. Assessing a gap in the biomedical literature: magnesium deficiency and neurologic disease. *Neuroscience Research Communications*, 15:1–9.

Smalheiser, N.R. and Swanson, D.R. 1996a. Indomethacin and Alzheimer's disease. *Neurology*, 46:583.

Smalheiser, N.R. and Swanson, D.R. 1996b. Linking estrogen to Alzheimer's disease: an informatics approach. *Neurology*, 47:809–810.

Smalheiser, N.R. and Swanson, D.R. 1998. Calcium-independent phospholipase A2 and schizophrenia. *Archives of General Psychiatry*, 55:752–753.

Swanson, D.R. 1986. Fish oil, Raynaud's syndrome, and undiscovered public knowledge. *Perspective in Biological Medicine*, 20:7–18.

Swanson, D.R. 1987. Two medical literatures that are logically but not bibliographically connected. *Journal of the American Society of Information Science*, 38:228–233.

Swanson, D.R. 1988. Migraine and magnesium: eleven neglected connections. *Perspectives in Biology and Medicine*, 31:526–557.

Swanson, D.R. 1990. Somatomedin C and arginine: implicit connections between mutually-isolated literatures. *Perspectives in Biology and Medicine*, 33:157–186.

Swanson, D.R. 1991. Complementary structures in disjoint science literatures, in A. Bookstein et al. (Eds.), *SIGIR91: Proceedings of the Fourteenth Annual International ACM/SIGIR Conference on Research and Development in Information Retrieval Chicago*, Oct 13–16, 1991, pp. 280–289. New York: Association for Computing Machinery.

Swanson, D.R. and Smalheiser, N.R. 1999. Implicit text linkages between Medline records: using Arrowsmith as an aid to scientific discovery. *Library Trends*, 48:48–59.

Weeber, M., Klein, H., de Jong-van den Berg, L.T.W., and Vos, R. 2001. Using concepts in literature-based discovery: simulating Swanson's Raynaud-fish oil and migraine-magnesium discoveries. *Journal of the American Society for Information Science and Technology*, 52(7):548–557.

Wilbanks, J. and Boyle, J. 2006. Introduction to Science Commons. Available at www.sciencecommons.org.

Wilkinson, N. 2002. Semantic Web use cases and case studies. World Wide Web Consortium (W3C) use cases. Available at http://www.w3c.org/2001/sw/sweo/Public/UseCases/Pfizer/.

Williams, A. 2005. Battling the data avalanche: a chemical data management solution for the start-up company. *Abstracts of Papers, 229th ACS National Meeting*, San Diego, CA, March 13–17, CINF-011, American Chemical Society.

Williams, A. and Yerin, A. 1999. The need for systematic naming software tools for exchange of chemical information. *Molecules,* 4(9):255–263. Available at http://www.mdpi.org/molecules/papers/40900255.pdf.

Zimmermann, M., Bui Thi, L.T., and Hofmann, M. 2005. Combating illiteracy in chemistry: towards computer-based chemical structure reconstruction. *ERCIM News,* 60.

Zimmermann, M. and Hofmann, M. 2007. Automated extraction of chemical information from chemical structure depictions. *Drug Discovery: Information Technology & Software Reports,* 12–15.

2 Chemical Information Mining: *A New Paradigm*

Debra L. Banville

CONTENTS

INTRODUCTION

Computer-assisted extraction or mining of chemical structural information from the literature requires special tools that address the various ways of encoding structures. Traditionally, in the literature, chemical structures are identified by textual names or images of structures. Chemical images of structures are in general very explicit and can convey a great deal of information to a chemist, but they cannot be read by computers (Ibison et al. 1993). To make these images machine-readable would involve a chemical image recognition capability as described in Chapter 4. This is a challenging area, and Chapter 4 provides the reader with an excellent background on what to expect when using these capabilities.

Textual names have the advantage of being machine-readable, but many chemical compound names are derived in the absence of structural information. Common or trivial names were frequently given to compounds based on their properties or methods of extraction (because their structural information was originally an unknown). For example, mandelic acid was extracted from bitter almonds, and its name is derived from *Mandel*, the German word for *almond* (Merck 1989).

Rule-based or systematic nomenclature (e.g., International Union of Pure and Applied Chemistry [IUPAC] or CAS nomenclature) is based on a set of linguistic rules that apply to its structure. For example, the IUPAC and CAS names for mandelic acid are 2-phenyl-2-hydroxyacetic acid and benzeneacetic acid, α-hydroxy-, respectively. The number of possible systematic names is practically endless. A small, but not nearly exhaustive, set of other systematic names includes phenylglycolic acid, phenylhydroxyacetic acid, (±)-α-hydroxybenzeneacetic acid, (±)-α-hydroxyphenylacetic acid, (±)-2-hydroxy-2-phenylethanoic acid, (±)-Mandelic acid, (RS)-Mandelic acid, DL-Amygdalic acid, DL-Hydroxy(phenyl)acetic acid, DL-mandelic acid, paramandelic acid, α-hydroxy-α-toluic acid, α-hydroxyphenylacetic acid, α-hydroxybenzeneacetic acid, 2-hydroxy-2-phenylacetic acid, 2-phenyl-2-hydroxyacetic acid, and 2-phenyl-glycolic acid. The application of these various linguistic rule sets and the multiple lexicons within each rule set makes conversion of a name to an actual structure very difficult for both computers and expert practitioners. Furthermore, chemical naming of compounds may define the substance but not its form. For example, glucose can exist in an open chain form or two possible closed chain forms (alpha-glucose and beta-glucose). It is not always clear what form or forms were intended when simply mentioning "glucose." In many cases the exact form or forms are not known. The implications of this during an automated name-to-structure conversion, are described in Chapter 3 along with a more detailed understanding of the chemical name-to-structure process itself.

The use of computer-readable formats for chemical structures has become a compelling need to capture this information in databases or to simply annotate documents (Degtyarenko et al. 2007). Annotating full text documents with these machine-readable forms can make documents easier to search, and the intended structure can be visualized in context. The development of computer-readable formats started in earnest around 1990 (Borkent et al. 1988; Weininger 1988; Contreras et al. 1990; Ibison et al. 1993).

The most commonly used identifiers today include line notation identifiers (e.g., Simplified Molecular Input Line Entry System [SMILES] and International Chemical Identifier [InChIs]), tabular identifiers (e.g., Molfile and Structure Definition [SD] file types), and portable mark-up language identifiers (e.g., Chemical Markup Language [CML] and FlexMol). Each identifier has its strengths and weaknesses as detailed in Chapter 5. Chapters 5 and 6 provide enough information to guide researchers in choosing the most appropriate formats for their individual use.

Chemical identifiers can be packaged inside documents in a variety of methods including eXensible Markup Language (XML; see Chapter 6 for details). XML has become so widely used in tagging documents with key information, and especially chemical information (e.g., CML), that we have dedicated Chapter 6 to this topic. These chemical structure tags, through XML, enrich the document and allow researchers to find documents tagged with their structures of interest and see how the compound is mentioned in the document. This contextual component can be a very simple and powerful research tool that paves the way for a new paradigm in chemical information mining of the literature, using text analytical tools such as chemical name entity recognition (NER) together with natural language processing (NLP) (Weizenbaum 1966; Jackson and Moulinier 2002). Chapter 7 provides clear coverage of this area.

Chemical NER can provide researchers with a very different experience when reading a tagged document. Some of these principles are captured in the Royal Society of Chemistry's Project Prospect, where key terms, such as compound names, are highlighted. Project Prospect is an excellent example of how access to information can be improved using a set of core noncommercial capabilities (Weininger 1988; Murray-Rust et al. 1997; Rupp et al. 2007; Corbett et al. 2007; Batchelor and Corbett 2007; Corbett and Murray-Rust 2006; Copestake et al. 2006, Smith et al. 2007; de Matos et al. 2006; Nic et al. 2002).

The ability to select a chemical name within the text, and view its structure, is starting to appear in a variety of tools and will most likely have a large impact in the area of patent searching. The application of this to patent documents has primarily come from the commercial sector from companies such as SureChem (http://www.surechem.org), InfoChem with IBM (http://infochem.de/en/mining/annotator.shtml), TEMIS with Elsevier MDL (http://www.temis.com), and Mpirics (http://www.mpirics.com/). This is a sample list, not meant to be exhaustive and not meant to be an endorsement of these products. An illustration of chemical NER of patents is shown in Figures 2.1a and 2.1b.

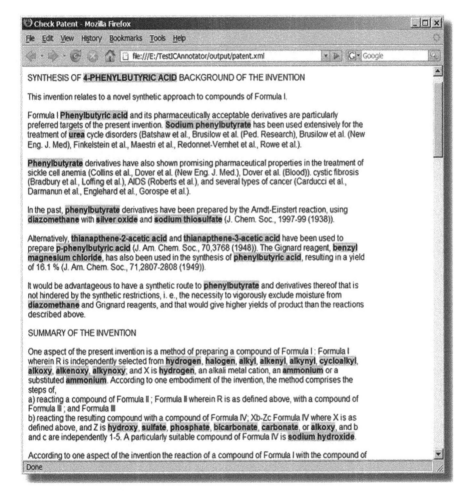

FIGURE 2.1A Screen shot showing examples of chemical NER patent capabilities from InfoChem's ChemAnnotator. "The extraction of chemically relevant entities from unstructured text sources is performed by software that recognizes and extracts systematic names, and trivial and trade names, as well as standard identifiers such as InChI's or CAS Registry Numbers. InfoChem is cooperating with leading companies in this area to utilize sophisticated, finely tuned software tools. The IBM Chemical Annotator is able to process English text files within seconds" (quoted from http://infochem.de/en/mining/annotator.shtml). (Note: This is not an endorsement of these vendors; it is only an illustration of chemical NER tools.)

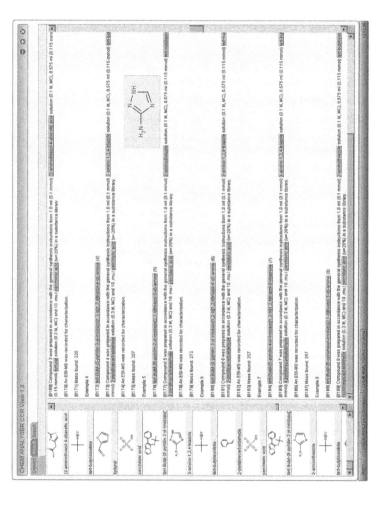

FIGURE 2.1B Screen shot showing examples of chemical NER patent capabilities from Mpiric's ChemAnalyser capability. "Patents contain a huge amount of unmined chemical information in text form. We have developed a Chemical Content Recognition (CCR) technology that employs multi-level error tolerant algorithms to recognise not only perfect chemical names but even misspelled names and artifacts due to suboptimal OCR. Based on this core technology, we have created CHEM ANALYSER that unveils chemical information and provides powerful research tools. With our novel hybrid view reading large chemistry-related patents becomes an easy and enjoyable task" (Dr. Michael Thormann, chief technology officer).

When layered onto chemical NER, NLP can provide readers with a richer experience. Natural language processing (NLP) technology does not understand human speech, but dissects language into the parts of speech such as nouns, verbs, and noun phrases. When combined with NER, NLP can assist the researcher in finding important relationships between chemical entities and their attributes. For example, in evaluating toxicity issues, NER and NLP should be able to highlight relationships between a compound of interest and a variety of liver toxicity biomarkers. The researcher still has to validate the relationships highlighted, but these types of capabilities can reduce the time required to find the relevant information within a large document set. Chapter 7 expands on these ideas.

Ultimately, the use of these capabilities has to enhance the ways in which researchers in both academics and industry work. Information overload is a major driver in this shifting paradigm, along with a variety of technological advances in other key areas. Recent developments in the Semantic Web are starting to bridge the gap in information overload. The techniques that are being used and developed to improve the ease of finding of information on the Internet are already having an impact on the mining of information within the literature. Chapter 8 presents a balanced view of this topic, sifting out the hype from the reality of the situation while highlighting the gap that must be bridged to create a Semantic Web ecosystem for the life sciences. Finally, Chapter 9 elaborates on the typical workflows necessary today for academic and industrial researchers needing to find the "right" information. This chapter outlines a road map for improvement that developers of these capabilities could use to effectively "close the loop."

REFERENCES

Batchelor, C.R. and Corbett, P.T. 2007. Semantic enrichment of journal articles using chimerical named entity recognition. *Proceedings of the ACL 2007*, demo and poster sessions, Prague, June 2007, pp. 45–48.

Borkent, J.H., Oukes, F., and Noordik, J.H. 1988. Chemical reaction searching compared in REACCS, SYNLIB and ORAC. *Journal of Chemical Information and Computer Science*, 28(3):148–150.

Contreras, M.L., Allendes, C., Alvarez, L.T., and Rozas, R. 1990. Computational perception and recognition of digitized molecular structures. *Journal of Chemical Information and Computer Science,* 30:302–307.

Copestake, P., Corbett, P., Murray-Rust, P., Rupp, C.J., Siddharthan, A. Teufel, S., and Waldron B. 2006. An architecture for language processing for scientific texts. *Proceedings of the UK e-Science Programme All Hands Meeting 2006 (AHM2006)*, Nottingham, UK.

Corbett, P., Batchelor, C., and Teufel, S. 2007. Annotation of chemical named entities. BioNLP 2007: Biological, Translational, and Clinical Language Processing, Prague, June 2007, pp. 57–64.

Corbett, P.T. and Murray-Rust, P. 2006. High-throughput identification of chemistry in life science texts, in *Computational Life Science II,* Lecture Notes in Computer Science series, Vol. 4216. Berlin/Heidelberg: Springer, pp. 107–118.

Degtyarenko, K., Ennis, M., and Garavelli, J.S. 2007. "Good annotation practice" for chemical data in biology. Workshop on Storage and Annotation of Reaction Kinetics Data, May 2007, Heidelberg, Germany. *In Silico Biology,* 7S1:06.

de Matos, P., Ennis, M., Darsow, M., Guedj, K., Degtyarenko, K., and Apweiler R. 2006. ChEBI—chemical entities of biological interest. *Nucleic Acids Research, Database Summary Paper*, 646.

Ibison, P., Jacquot, M., Kam, F., Neville, A.G., Simpson, R.W., Tonnelier, C., Venczel, T., and Johnson, A.P. 1993. Chemical literature data extraction: the CliDE project. *Journal of Chemical Information and Computer Science,* 33(3):338–344

Jackson, P. and Moulinier, I. 2002. *Natural Language Processing for Online Applications: Text Retrieval, Extraction, and Categorization,* Vol. 5. Philadelphia: John Benjamin's Publishing Co.

Merck & Co. 1989. *The Merck Index,* 11th ed. Rahway, NJ: Author, p. 5599.

Murray-Rust, P., Rzepa, H.S., and Whitaker, B.J. 1997. The World Wide Web as a chemical information tool. *Chemical Society Reviews,* 26:1–10.

Nic, M., Jirat, J., and Kosata, B. 2002. *Compendium of Chemical Terminology* (also known as the *IUPAC Gold Book*). Prague: ICT Press. Available at http://goldbook.iupac.org/index.html.

Rupp, C.J., Copestake, A., Corbett, P., and Waldron B. (2007). Integrating general-purpose and domain-specific components in the analysis of scientific text. *Proceedings of the UK e-Science Programme All Hands Meeting (AHM2007),* Nottingham, UK. Available athttp://www.allhands.org.uk/2007/proceedings/papers/860.pdf.

Smith, B., Ashburne, M., Rosse, C., Bard, J., Bug, W., Ceusters, W. Goldberg, L.J., Eilbeck, K., Ireland, A., Mungall, C.J., Leontis, N., Rocca-Serra, P., Ruttenberg, A., Sansone, S.-A., Scheuermann, R.H., Shah, N., Whetzel, P.L., and Lewis, S. 2007. The OBO Foundry: coordinated evolution of ontologies to support biomedical data integration. *Nature Biotechnology,* 25:1251–1255.

Weininger, D. 1988. SMILES: a chemical language and information system. *Journal of Chemical Information and Computer Sciences,* 28(1):31–36.

Weizenbaum, J. 1966. Eliza: A computer program for the study of natural language communication between man and machine. *Communications of the ACM,* 9(1):36–45.

Part II

Chemical Semantics

3 Automated Identification and Conversion of Chemical Names to Structure-Searchable Information

Antony J. Williams and Andrey Yerin

CONTENTS

INTRODUCTION

Chemical names have been in use as textual labels for chemical moieties even since the days of the alchemist. With increasing understanding of chemistry and the graphical representation of chemical structures came the need for an agreed upon language of communication between scientists. Eventually, systematic nomenclature was established and then extended as deeper knowledge and understanding of molecular structures grew. One would hope for a single agreed upon international standard for systematic nomenclature adopted and understood by all chemists. Despite the efforts of the IUPAC,[5] such an ideal still does not exist, exists in many variations, has changed over time, can be organizationally specific, is multilingual, and is certainly complex enough that most chemists would struggle with even the most general heterocyclic compounds. The application of nomenclature by scientists of different skill levels is far from pure, and chemical names for a single species are heterogeneous. This does not bode well for clear communication in chemistry.

Chemical nomenclature is a specific language for communication between people with an understanding of chemistry. The language facilitates the generation of chemical names that are both pronounceable and recognizable in speech. The ability to communicate via systematic names collapses fairly quickly based on the complexity of the chemical structure and the associated name. Simple and short names are easily interpreted, but in general most systematic names are rather long, complex, and include nonlinguistic components such as locants and descriptors made up of obscure numbers and letters. A chemical nomenclature system must continuously follow the increasing complexity and diversity of chemical structures as new chemistries are pursued. The majority of chemical names are rather complex, and a chemist needs a reasonable knowledge of the nomenclature rules to interpret a chemical name and convert it back to a graphical structure representation. Chemical nomenclature rules and recommendations for the IUPAC are now captured online in a series of volumes with several thousand pages.[6]

Despite the limitations and challenges associated with chemical names, graphical chemical structure representations, on the other hand, can easily be interpreted by humans even with the most rudimentary chemistry knowledge. Chemical structure representations were in use well before the advent of software programs for the generation of such figures. Structure-drawing software was developed to provide a way to store, transfer, and homogenize molecular structure representations. The ability to both represent and transfer chemical structures electronically provided a significant boost to communication between chemists, and structure images became the preferred medium for human recognition. Despite the availability of software tools for the graphical representation of chemical structures, chemical names, labels, and abbreviations are still required for us to converse. They remain as valuable terms of communication in patents and publications and are essential to the process of chemical registration for a number of bodies. The generation of appropriate systematic nomenclature remains a challenge to even the most skilled chemist, but because systematic nomenclature is rules-based, the development of software tools to speed the process has been possible. The opposite is also true, whereby the conversion of systematic names to the original chemical structures remains just as much of a challenge. By providing software tools for the conversion of differing chemical nomenclatures into universally recognized chemical structures, chemists can more easily review the chemical structure of interest, and the data can be migrated to database technologies. This facilitates the integration of disparate forms of chemical information with the intention of enabling the discovery process.

There are numerous sources of chemical names. Commonly, chemical databases did not include chemical structures but were made up of lists of chemical names.

Nowadays, thanks to the availability, cost, and ease-of use of chemical structure data-bases, many of these "text databases" have been converted into a structure format, and most chemical databases are now structure searchable. A simple search of the Internet will show that many databases still lack chemical structures and therefore are not searchable by structure in the original format, for example, an online HTML page. These pages, however, can contain valuable information and, with the application of the appropriate name-to-structure (N2S) conversion tools can be made searchable.

Electronic documents exist in a plethora of formats, the most common being Microsoft Word, portable document format (PDF), and web-based HTML for-mats, as well as a number of others. Electronic documents in general do not embed information regarding chemical structures, but do include chemical names that are extractable. It is likely that nearly all modern documents of interest to chemists are now available in electronic format. Published both before and after the early stages of computerization, such documents might be considered lost for chemical informa-tion. However, scanning and optical character recognition[7] (OCR) into electronic files provides a means for conversion by software tools. Of course, even without such tools, scientists commonly read print documents and manually convert the chemical names to structures. It should be noted that it is also possible to identify chemical images and convert them to structure-searchable information using optical structure recognition (OSR). This is discussed in detail in Chapter 4 of this book.

The conversion of chemical names and identifiers into appropriate chemical structure representations offers the ideal path for chemists and organizations to mine chemical information. Because chemical names are not unique and a multitude of labels can map to a single chemical entity, the facile conversion of alphanumeric text identifiers to a connection table representation enables superior data capture, representation, indexing, and mining. The industry's need to mine more information from both the historical corpus as well as new sources is obvious, and a number of researchers have initiated research into the domain of chemical identifier text min-ing and conversion. Multiple efforts have been made in the field of bioinformatics research,[8] and, while interesting as a parallel, in this chapter we will focus the efforts to extract and convert identifiers related to chemical entities rather than, for example, genes, enzymes, or proteins.

Our intention in this chapter is to examine the challenges of extracting identifiers from chemistry-related documents and the conversion of those identifiers into chem-ical structures. The authors of this work each have well over a decade of experience in chemical structure representation and systematic nomenclature. We have been deeply involved in the development of software algorithms and software for the gen-eration of systematic names and the conversion of chemical identifiers into chemical structures.[9] Although we have our own biases concerning approaches to the problem of N2S conversion, we have done our utmost to be objective in our review of the subject and comparison of approaches and performance.

EXISTING STRUCTURE MINING TOOLS AND PROJECTS

It is likely that ever since systematic nomenclature was introduced, chemists have wished for a simple way to convert a systematic name to a graphical representation of the associated structure. A number of organizations have built business models around the extraction and conversion of chemical names from different materials (e.g., publications, patents, and chemical vendor catalogs) to build up a central repository of chemical structures and links to associated materials. The Chemical Abstracts Service (CAS) is recognized as the premier database and presently contains over 33 million compounds.[10] Other offerings include those of Beilstein,[11] Symyx[12] (previously MDL), Infochem,[13] and VINITI.[14] These organizations manually curate, nowadays with the assistance of software tools, chemical structures and reactions from the respective publications and documents.

The delivery of new chemical entities of commercial value can clearly be constrained by the coverage of patent space. Chemical structure databases linked to patents are available (e.g., CAS,[2] Elsevier,[15] and Derwent[16]) and deliver high value to their users. Some of these organizations utilize both text-mining and N2S conversion tools prior to manual examination of the data. Two free-access services utilizing text mining and conversion of chemical names to structures are those of SureChem[17] and IBM.[18]

Both approaches use proprietary entity extraction tools developed and customized specifically for the recognition of chemical names.[19,20] The chemistry-specific entity extractors use a combination of heuristics for systematic names and authority files for entities that are less amenable to rules-based recognition, specifically drug and chemical trade names. During the extraction and conversion processes, chemical entities are run through one or more N2S conversion tools to generate chemical structure data. A set of postprocessing routines are applied to remove spelling and formatting errors that often cause N2S conversion failure, but experience has shown that due to the poor quality of many chemical names in patents and other text sources, not all of the names can be converted by commercially available tools.

SureChem offers a free-access website for searching the world patent literature via text-based or structure and substructure searching,[17] as well as commercial offerings based on the same chemical patent data (for example, they supply the data in formats to allow importing of the data into organizational databases). They have utilized a series of N2S conversion tools under their system, supplied by three commercial entities: ACD/Labs,[21] CambridgeSoft,[22] and Openeye.[23] They provide ongoing updates of the patent literature within 24 hours of release to the public and update their homepage accordingly with the latest statistics of extracted chemical names, details regarding each of the patent classes, and the number of unique structures extracted to date. SureChem reports the extraction of over half a billion chemical structures[17] from various patent-granting bodies, and these have been de-duplicated to almost 9 million unique structures. They offer online access to various forms of patent literature including U.S.- and European-granted applications as well as WO/PCT documents[24] and Medline.[25] All of these sources are updated within a day of release of the updates from the patent offices to SureChem. N2S conversion results vary among the patent databases due to different levels of original text quality among the patent issuing authorities. SureChem reports[26] that in their latest database build they observed improvements of as much as 20% in N2S conversion rates following application of new postprocessing heuristics and expect further incremental increases in future builds.

IBM[18] also has a free-access online demonstration system for patent searching via text or structure and substructure and presently exposes data extracted from U.S. patents (1976–2005) and patent applications (2003–2005). The work has been described in detail by Boyer et al.,[27] and a brief overview of the technology is provided on the website.[28] They report using the CambridgeSoft Name=Struct[22] algorithms for their work. The IBM team has also analyzed both granted patents and patent applications to the present day for all sources listed above, but these data are not yet exposed at their website, and the exposed data are limited to U.S. Patent and Trademark Office patents and Medline articles issued up to 2005. IBM reports the extraction of over 4.1 million unique chemical structures. Caution should be used when comparing unique chemical structures reported by SureChem and IBM, as the methods of de-duplication are not necessarily comparable and are not reported in detail. At present, SureChem is the most mature free-access online service, updated on a regular basis and covering a number of patent-granting bodies.

Accelrys[29] has also developed text analytics capabilities for the purpose of extracting and converting chemical names. Using their Scitegic pipelining tools as the platform, they developed the ChemMining[30] chemical text mining and conversion system. This software uses text-mining algorithms to extract chemical names and then feeds these to one or more of the commercial N2S conversion algorithms licensed by the user. After one or more documents are processed, a report is created showing the examined document(s) highlighted with all of the found structures as live chemistry objects.

Murray-Rust et al.[31,32] have examined the challenges associated with mining data from text and have encouraged the adoption of appropriate architectures, molecular identifiers, and a shift toward more open data to facilitate information exchange in the sciences. They have appropriately espoused the virtues of their OSCAR system,[32] a chemical data checker in an Open XML architecture, in terms of its benefits to authors, publishers, and readers. In this work, compounds were identified by connection table links to open resources such as PubChem.[33] Originally a part of the OSCAR system, OPSIN[32,34] (Open Parser for Systematic Identification of Nomenclature) has been released as an Open Source Java library for parsing IUPAC nomenclature. OPSIN is limited to the decoding of basic IUPAC nomenclature but can handle bicyclic systems and saturated heterocycles. OPSIN does not deal with stereochemistry, organometallics, or many other expected domains of nomenclature, but because the source code is open, it is hoped that this work can provide a good foundation technology for others to enhance and develop.

TEMIS and Elsevier MDL[35] worked together[36] to develop the Chemical Entity Relationship Skill Cartridge to identify and extract chemical information from text documents. The software identifies chemical compound names, chemical classes, and molecular formulas and then translates them into chemical structures. They use an N2S translation service to match textual information with proprietary chemical libraries and provide a unique fingerprint for de-duplication purposes. The cartridge integrates chemical name recognition software developed and used by Elsevier MDL to identify chemical names and extract reaction schemes from scientific literature and patents. This software was proven for more than two years in the production of the MDL Patent Chemistry Database, including processing a backlog of more than 20 years of patents. Unfortunately, these authors cannot locate any further details regarding the software or performance. Research into text-mining continues to expand, and a national center of text mining, with a focus on the sciences, has been founded in the United Kingdom.[37]

The projects outlined above all focus on the extraction of chemical identifiers from text, but there is a clear dependence on the N2S conversion algorithms for the overall output of the various approaches. The remainder of this chapter will review the challenges associated with the development of N2S algorithms and how these can be addressed.

THE GENERAL APPROACH TO MINING CHEMICAL STRUCTURES IN CHEMICAL TEXTS

The scheme by which chemical structures are mined from chemical documents is shown in Figure 3.1. The greatest hurdle associated with successful mining of chemical structures via chemical N2S conversion is the quality and complexity of the chemical names themselves. Thus, a significant part of this chapter is devoted to the consideration of the quality of names and its contributions to the procedure of conversion.

Text Recognition in Images: OCR of Chemical Texts

Starting from the very beginning of OCR technologies, a huge amount of resources was invested in the development of computer-based systems. For general language-based texts this problem has been efficiently solved, and the success rate of recognition is higher than 99% for Latin-script texts.[38] The basic challenges of OCR have been reviewed elsewhere and will not be repeated here.[39] Although OCR can efficiently handle generic text, it experiences fairly significant limitations in the

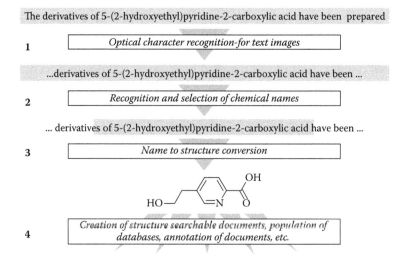

FIGURE 3.1   General scheme for mining chemical structures from text.

recognition of chemical names. In the same way that general OCR programs use language-specific dictionaries to assist in recognizing text, a chemically intelligent OCR program needs to use a dictionary of appropriate chemical text fragments and use a series of specific algorithms to recognize chemical names. Figure 3.2 illustrates the recognition of chemical name images captured with different settings. A standard software package was utilized for these test procedures.[40] Each of the examples shows the graphical image of the chemical name as well as that extracted by the software.

Although it is clear that recognition of this example can be easily improved by enhanced resolution of the initial image, this example is given to demonstrate the most common problems associated with chemical name recognition and possible errors introduced in chemical names. The problems include:

- Superscript and subscript recognition, especially in combination with italics
- Introduction of additional spaces, often instead of paragraphs
- Lost spaces, mainly at line breaks
- Dashes often lost or mistaken as hyphens
- Incorrect recognition of punctuation marks (e.g., comma versus period)
- Misinterpretation of enclosing marks
- Incorrect recognition of some letters and numbers (e.g., l, i, 1)
- Lost formatting (e.g., normal text versus sub- or superscripted characters)

As a result of these considerations, we can conclude that OCR of chemical names can be improved by:

- Utilizing higher-resolution text images
- Usage of chemical dictionaries
- Modification of OCR algorithms for chemical name recognition and specifically for retaining dashes and avoiding added spaces

(2*S*,5*R*,6*R*)-6-{[(3-aminotricyclo[3.3.1.1^{3,7}]dec-1-yl)acetyl] amino}-3,3-dimethyl-7-oxo-4-thia-1-azabicyclo[3.2.0]heptane-2-carboxylic acid

▼

(2S,5R,6**Rp6**-{[(**3-aminotricyclo[3.3.1.1??]**dec-1-yl)acetyl**J**amino}**3,3** dimethyl-7-oxo-4-thia-1-azabicyclo[3.2.0]heptane-2-carbo**o**cylic acid

(2*S*,5*R*,6*R*)-6-{[(3-aminotricyclo[3.3.1.1^{3,7}]dec-1-yl)acetyl] amino}-3,3-dimethyl-7-oxo-4-thia-1-azabicyclo[3.2.0]heptane-2-carboxylic acid

▼

(25,5R,6R)-6-[[(3-aminotricyclo[3.3.1.1V]dec-1-yl)acetyl]amino}**3,3** dimethyl-7-oxo-4-I**hia**-1-aza**6**icy**d**o[**320**]heptane-2-car**6**oxylic acid

FIGURE 3.2 Problems with character recognition in chemical names.

CHEMICAL NAME SELECTION AND EXTRACTION

When text analysis is required as a result of either OCR conversion or simply from direct electronic formats, the selection or recognition of chemical names becomes the challenge. As stated earlier, the nature of chemical names can vary widely and be represented either by single words or as sets of grammatically linked words. Another difficulty is that text within a chemistry context can include terms derived from chemical names that serve as verbs, adjectives, or plural forms describing processes, chemical relations, or groups of chemical substances. For example, in the phrase "acetylation of isomeric diethylnaphthyridines with acetic anhydride," only one distinct chemical name can be selected ("acetic anhydride"), though clearly the conversion of "diethylnaphthyridines" as a class of compounds could lead a reader to a text of interest.

The first publications in this area were from the 1980s and 1990s.[41,42] This area of research now uses the general principles of natural language processing (NLP) and, specifically, named entity extraction (NER) enhanced with specific developments for chemical and biochemical name recognition.[43,44] Chapter 7 of this book is devoted to NLP and NER approaches applied to the extraction of chemical information, and we will not discuss these approaches in more detail here.

The specific problems and potential solutions associated with chemical name recognition have been reviewed in a recent work describing the OSCAR3 software.[32] The general approach is the recognition of chemistry-related terms whereby chemical names are identified by the appropriate algorithms. Chemical name identification uses several steps and procedures that may include:

- Splitting words with common separators such as spaces and punctuation marks with spaces according to natural language and chemical name rules
- Recognition of chemical words using dictionaries of chemical lexemes
- Syntax and semantics analysis of relationships between words to recognize chemical names that include spaces

Following the chemical name recognition process, annotated documents are created with specific tags to provide a reference to the part of the document where the specific chemical is mentioned. The extracted chemical names are then provided as inputs to the N2S algorithms and form the basis of the next section of this work.

GENERATING CHEMICAL STRUCTURES FROM CHEMICAL NAMES

ALGORITHMIC N2S CONVERSION AND RELATED SOFTWARE APPLICATIONS

The first publication about the computer translation of chemical names was published by Garfield in 1961. In that article, he described the conversion of names into chemical formulas and initiated the path toward N2S algorithm development.[45] Developments in 1967 at CAS provided internal procedures for the automatic conversion of CAS names into chemical diagrams.[46,47] The first commercially available software program was CambridgeSoft's Name=Struct released in 1999,[48] now patented,[49] which was followed shortly by ACD/Labs' ACD/Name to Structure product released in 2000.[50] Two more commercial products are available: ChemInnovation's NameExpert[51] and

OpenEye's Lexichem,[23] and ChemAxon[52] has announced the imminent release of their own product early in 2008. As mentioned earlier, an Open Source Java library for the interpretation of IUPAC systematic names,[34] OPSIN, has also been made available. In this chapter, most examples are based on Name=Struct and ACD/Name to Structure. We judge these programs to currently be the most advanced products in this area, but all considerations are general in nature and relevant to all of the conversion routines presently existing or still under development.

The vision for all N2S conversion algorithms is likely consistent. Convert as many chemical names as possible to the correct chemical structures. Whereas this is the general target, the approaches to arrive there can differ. ACD/Labs have maintained an approach of caution in terms of name conversion, initially focusing only on the translation of fully systematic names, controlling ambiguity to as high a level as possible yet supporting the conversion of trivial names using a dictionary lookup. CambridgeSoft has approached the problem with the intention of converting as many names as possible and being fairly neutral in terms of name format and strict systematic nomenclature format. For many test comparisons, both approaches have their failings. ACD/Labs' product sometimes fails to successfully convert names, yet CambridgeSoft commonly converts a much larger proportion of the test set but with more inappropriate conversions. Many of the larger companies have chosen to support both approaches, licensing both tools and performing intersecting comparisons and examining the results outside of the intersection for appropriateness. This approach has been taken by the groups analyzing the patent literature as discussed earlier. SureChem uses three N2S products for their work as discussed earlier.

GENERAL SCHEME OF NAME TO STRUCTURE CONVERSION

The conversion of chemical names into chemical structures can be represented as two intersecting schemes: utilizing a lookup dictionary and using syntax analysis. A combination of these two approaches is definitely needed for the analysis of chemical names in the real world.

Figure 3.3 illustrates the simplest approach of using lookup tables. In this approach the N2S engine utilizes the relationship between a large database of chemical names and the corresponding chemical structures.

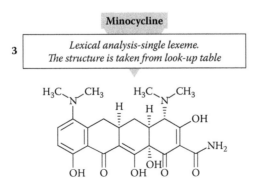

FIGURE 3.3   Single-step conversion of trivial name.

The rather restricted nature of this approach is obvious: The potential number of chemical structures and their associated chemical names is very large and cannot be included in a computer program of a reasonable size. Clearly, significant resources would be needed to create such a database of names and structures and keep it updated and distributed to users at the appropriate pace of chemical development. When the diversity of name formatting resulting from human intervention is taken into consideration, then this factor alone will make N2S conversion essentially intractable. A lookup table approach is nevertheless very useful, and InfoChem utilizes their in-house IC_{N2S} program for the purpose of chemical structure mining from texts using an internal file of 27 million names.[53] A lookup algorithm and associated databases are unavoidable for the treatment of trivial names and other structure identifiers such as registry numbers.

For the conversion of systematic names, a more powerful and flexible approach must be based on the parsing of the chemical names and the application of syntax analysis. Figure 3.4 illustrates the principle steps of this procedure. The first step in the process, lexical analysis, splits the whole chemical name into a series of name fragments, known as lexemes, that have structural or grammatical meaning. Also

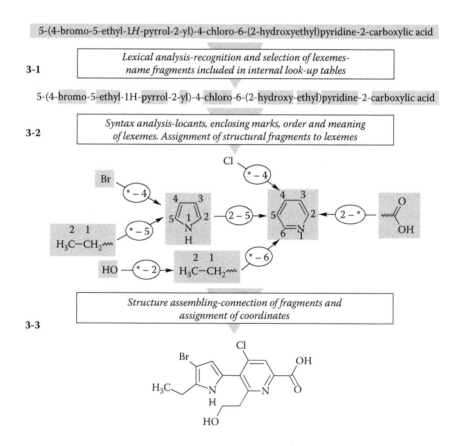

FIGURE 3.4   General steps of conversion of an unambiguous systematic name.

split out from the name are the locants, the enclosing marks, and the punctuation marks. If any part cannot be recognized by the program, then structure generation will normally fail or an attempt to continue generation by applying a rules-based spelling correction or ignoring a part of the input name can be performed. The lexical dictionary used at this stage is related to that described earlier to find the chemical names in the text.

The second step shown in the figure is the syntax analysis of the chemical name. At this stage the chemical name is analyzed according to chemical nomenclature grammar, each fragment is assigned its structural meaning, and attempts are made to derive a connection between the various structural fragments. In the simplest case of an unambiguous systematic name, all name parts can be interpreted in only one way, allowing the determination of a single chemical structure. This step is the primary component of an N2S engine. Many challenges and problems are associated with this engine, and these are discussed below for specific chemical names.

During the last step, all structural name fragments are assembled into a chemical structure, and atom coordinates are assigned to provide an attractive representation of the chemical structure for storage or exporting into various chemical formats including line notations, such as InChI (International Chemical Identifier) and SMILES (Simplified Molecular Line Entry System). The basic principles and problems of N2S conversion have been discussed previously by Brecher[54] in his description of the CambridgeSoft Name=Struct program. We will discuss further challenges of N2S conversion concerning specific types of chemical names in relation to the mining of chemical structures from texts.

CONVERSION OF TRIVIAL NAMES

As illustrated in Figure 3.1, the simplest N2S engine may be fully based on a lookup table and does not require the parsing of chemical names. As discussed above, although it is necessary to have large dictionaries of chemical names and structures, this approach is unavoidable for the conversion of names and structure identifiers where parsing cannot help in the process of structure generation. Such an algorithm can be used to convert trivial, trade, and retained names together with registry numbers such as CAS, EINECS, and vendor catalog numbers.

One important aspect of this approach deserves mention: The support of stereoisomerism requires caution. In many cases in the literature and in many databases, a specific stereoisomer is represented without definition of the configurations, and the specific stereoisomer is simply implied. Figure 3.5 shows several examples of such cases.

The structures shown in Figure 3.5 can give rise to 32, 2,048, and 512 different stereoisomers, respectively, and do not accurately represent the chemical names displayed below. It is very common that the representation of stereochemistry for both amino acids and steroids is not reported in publications. Caution must be used with the generation of the N2S engine structure dictionary from representations that omit stereo configurations, for example, from nonstereo SMILES notation.

In most cases, the N2S conversion of indivisible or elementary identifiers is safe, and the quality of conversion depends only on the internal dictionary quality. One

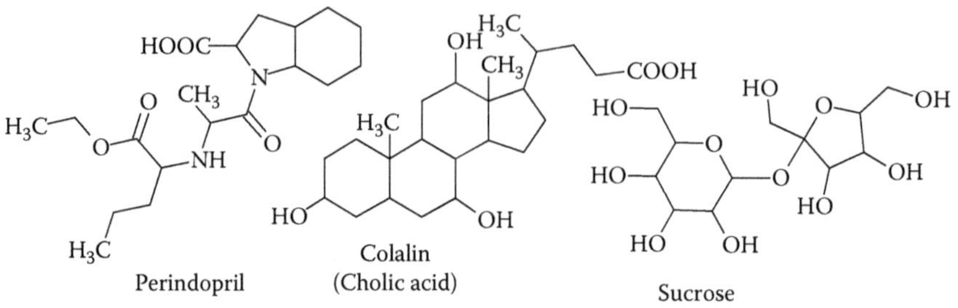

FIGURE 3.5   Stereoisomers represented without indication of configurations.

important exception is that of chemical abbreviations. Although they can be treated as "trivial names," they are very context dependent and highly ambiguous because such a limited number of letters cannot be treated as a unique identifier.

Figure 3.6 shows 12 structures that may correspond to the abbreviation "DPA." Six of them can be output by the ACD/Name to Structure software package, and six more were found by browsing the Internet. Note that even a specific context cannot guarantee an exact meaning. For example, both structures 3 and 8 were found in publications about coordination compounds. In general, chemical abbreviations are *not* unique and can rarely be distinguished from other trivial names except for the rather weak criterion that all letters are capitalized. We can conclude that conversion of any trivial name shorter than about five or six characters is not safe. A few rarer exceptions do exist, but this is a very short list. Examples include reserved abbreviations such as those for dimethyl sulfoxide (DMSO) and ethylenediaminetetraacetic acid, EDTA.

Conversion of Systematic Names

The lexical and syntax-based analysis of systematic names illustrated in Figure 3.2 depends directly on the algorithms underlying the name conversion engine. The set of lexemes that can be recognized by an algorithm are a critical characteristic of the program because it defines what type of names can be treated. However, the number of elementary lexemes is not the defining limitation of the program. The integration of the appropriate set of lexemes with the appropriate treatments for handling complex nomenclature grammar are superior to an extended set of lexemes. For example, the treatment of all fused system names requires the support of specific nomenclature grammar and approximately 100 specific lexemes. This approach is far more powerful than the support of 1,000 fused system names represented as elementary lexemes such as furo[3,2-b]pyridine, cyclobuta[a]naphthalene, and so on.

Chemical nomenclature has a very large number of specific procedures to create chemical names, and many of these are not easily amenable to algorithmic representation, requiring significant investments in both development and validation time to develop automated procedures. Software developers of N2S engines prefer to support just the basic operations for conversion, at least at the early stages of development.

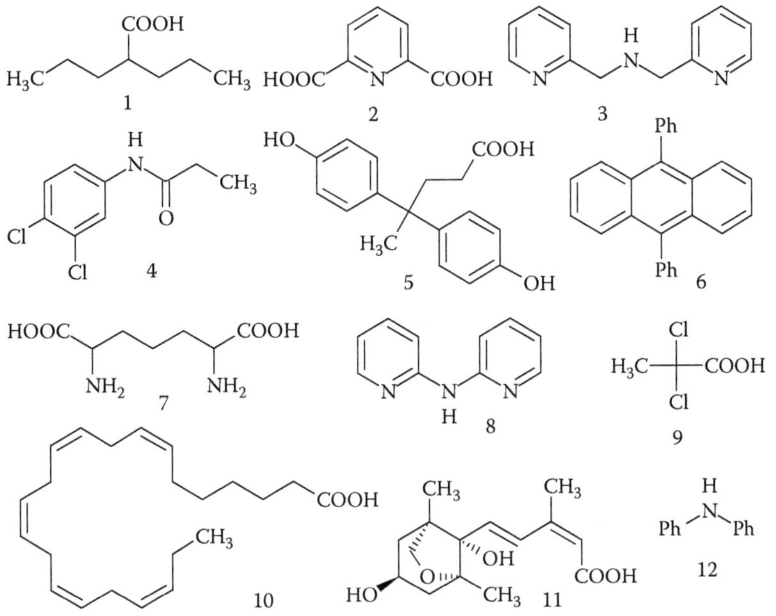

1 DiPropylAcetic acid   7 2,6-DiaminoPimelic Acid

2 DiPicolinic Acid   8 Di(2-Pyridyl)Amine

3 Di(2-Picolyl)Amine   9 DichloroPropionic Acid

4 3',4'-DichloroPropionAnilide 10 DocosaPentaenoic Acid

5 DiPhenolic Acid   11 DihydroPhaseic Acid

6 9,10-DiPhenylAnthracene   12 DiPhenylAmine

FIGURE 3.6 Twelve structures that may correspond to the abbreviation "DPA." The letters used in the abbreviation are bold and capitalized.

One of the largest challenges is that many chemical names, even when generated appropriately and without errors, are created according to different nomenclature systems. Specifically, the two most common nomenclature systems, those of the IUPAC and CAS, have many differences and can lead to potential ambiguity of the names. The situation becomes even more complex when we take into account the fact that chemical names have mutated through history with the development of the nomenclature systems and so, for example, many chemical texts follow old nomenclature procedures, thereby significantly expanding the number of nomenclature operations requiring support.

The conversion of systematic names to their chemical structures is a time-consuming, skill-intensive process, and is not a minor undertaking. Such a project is guaranteed to take many years of development to cover the most important nomenclature operations.

QUALITY OF PUBLISHED CHEMICAL NAMES

The main problem of name conversion is the rather low quality of published systematic names. It may be considered one of the reasons for the appearance of N2S programs. The paper describing CambridgeSoft's Name=Struct program has a very symbolic title: "Name=Struct: A Practical Approach to the Sorry State of Real-Life Chemical Nomenclature."[54] Most chemists have limited nomenclature knowledge, so resolving chemical names of fairly nominal complexity is a nontrivial task for them. The reverse is also true: The generation of systematic names for complex chemical structures can be a challenge, and as a result there has been a proliferation of incorrect structure–name pairs not only on the Internet but also in peer-reviewed publications. A recent review of systematic nomenclature of chemicals on Wikipedia by one of our authors (AJW) demonstrated significant gaps in quality, to the point where the names represented very different structures than those discussed on the Wikipedia pages. The quality of published systematic names is rather low, and this is true not only of publications but also of patents. In a recent paper, Eller[55] randomly selected about 300 names of organic chemicals cited to be systematic in nature. The names were extracted from four chemical journals and analyzed and compared to the corresponding names generated by a number of systematic nomenclature-generation software packages. The results of this comparison are given in Table 3.1.

Software for generating a systematic name from a structure has been available for well over a decade. Whether the issue is one of access to software or trust that software can produce high-quality systematic nomenclature, it is clear that papers still contain far too many errors in their systematic names. The data in Table 3.1 reflect the situation in 2006. Although this is not exactly a statistical sampling of data (only 300 names from four journals), the data suggest that about a quarter of published chemical names do not accurately represent the associated structures. There are two specific issues: (1) The chemical name does represent the structure and *can* be converted back to the intended structure, but the name does not follow systematic nomenclature guidelines; (2) the chemical name does not represent the structure and when converted generates a *different* structure from that originally intended. The data in the table clearly demonstrate that algorithmically generated names are of dramatically higher quality and reliability than manually generated names and that wider adoption of software programs for this purpose will significantly improve the quality of published nomenclature. The barriers to this shift are likely threefold:

TABLE 3.1

Comparison of Computer-Generated Names with Published Names (Results of Analysis of 303 Systematic Names)

	Unambiguous		Intolerable		No Name	
Published names	74%	224	26%	79		
AutoNom 2000	86%	260	1%	3	13%	40
ChemDraw 10.0	88%	267	1%	2	11%	34
ACD/Name 9.0	99%	300	1%	3	0%	0

awareness of the availability of such software applications, price and technology barriers to accessing such applications, and trust in the ability of the software to produce an appropriate systematic name. Attention must be given to improved generation of systematic nomenclature as soon as possible because the proliferation of poor quality and the contamination of the public records can now occur at an outstanding rate with new software platforms.

Thielemann[56] recently commented that the number of mistakes in systematic names is far higher than that of trivial names. He provided examples as a result of his examination of patents regarding the cholesterol-lowering drug Simvastatin. He observed that out of 141 patents examined, not one contained the correct IUPAC name of Simvastatin. He also pointed out what the correct IUPAC name, in his opinion, was: 6(R)-[2-[8(S)-(2,2-dimethylbutyryloxy)-2(S),6(R)-dimethyl-1,2,6,7,8,8a(R)-hexahydronaphthyl]-1(S)ethyl]-4(R)-hydroxy-3,4,5,6-tetrahydro-2H-pyran-2-one. Unfortunately this "correct name" is far from appropriate according to IUPAC rules, primarily due to the incorrect citation of stereodescriptors. Neither the CambridgeSoft Name=Struct nor the ACD/Labs Name to Structure software can convert the systematic name suggested by Thielemann back to the original Simvastatin chemical structure. In our judgment none of the commercially available N2S conversion algorithms can convert this name to the structure. The structure of Simvastatin with an appropriate IUPAC name is given in Figure 3.7.

This example demonstrates that one of the main challenges for an N2S conversion algorithm applied to data mining is the conversion of chemical names that are not strictly systematic, are ambiguous, or include typographical errors or misprints.

AMBIGUOUS SYSTEMATIC NAMES

It is not difficult to identify many ambiguities in chemical names in chemical catalogs, publications, patents, and Internet pages. Even the simplest structures can be given ambiguous names and cause confusion. Figure 3.8 shows a series of examples of names with missing locants or parentheses that often, but not necessarily, lead to name ambiguity.

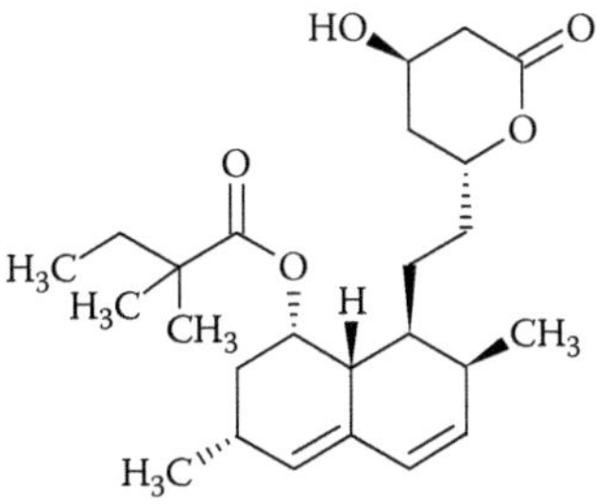

(1S,3R,7S,8S,8aR)-8-{2-[(2R,4R)-4-hydroxy-6-oxotetrahydro-2H-pyran-2-yl]ethyl}-3,7-dimethyl-1,2,3,7,8,8a-hexahydronaphthalen-1-yl 2,2-dimethylbutanoate

FIGURE 3.7   Chemical structure and IUPAC name of Simvastatin.

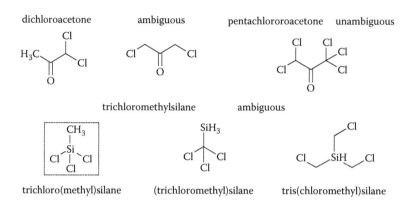

FIGURE 3.8 Potential ambiguity of names with missing locants or parentheses.

It should be noted that the name *trichloromethylsilane* is the correct CAS name for the framed structure that provides legal status to some ambiguous names. A more complex example of ambiguity introduced by missing parentheses is shown in Figure 3.9. In this case the recognition of ambiguity requires support of a specific nomenclature procedure: functional modification of trivial acid names.

In an example such as this, there are a number of ways to proceed: (1) Convert the name to a single acceptable structure matching the ambiguous name; (2) do not convert the name to a structure but fail because of the ambiguous nature of the name; (3) convert the name to all possible structures to demonstrate potential ambiguity. For the example in Figure 3.9, the commercial software providers take different paths. ACD/Name to Structure generates two structures for this name, whereas CambridgeSoft Name=Struct outputs only the second structure, because it is the most probable match, given that the correct systematic name of the first structure is **4-(methylthio)benzoic acid**. For the >550 hits returned by a search in Google most, but not all, refer to the first structure.

A similar example is "4-methylthiophenol." This name also allows the generation of two structures, but here the situation is reversed and most cases refer to 4-methyl(thiophenol) (or 4-methylbenzenethiol according to current naming conventions).

This short overview with simple examples provides evidence for the need of warnings regarding ambiguity in names. Clearly, the more complex a chemical structure is, the more potential there is for miscommunication. It is our belief that the recognition and reporting of ambiguities in chemical names and the associated structures generated by software programs must be implemented as part of any N2S engine to ensure some level of caution to provide reliable results.

AMBIGUOUS VERSUS TRIVIAL NAMES

One of the primary issues with systematic nomenclature is that some names can appear systematic in nature but, in fact, are not. They can have the expected structure of a chemical name generated according to a rules-based system but are false systematic names, at least in their specific context. When the N2S conversion algorithms

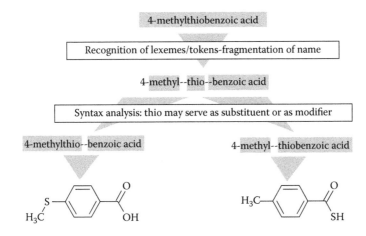

FIGURE 3.9 Different meanings of "thio" in an ambiguous name.

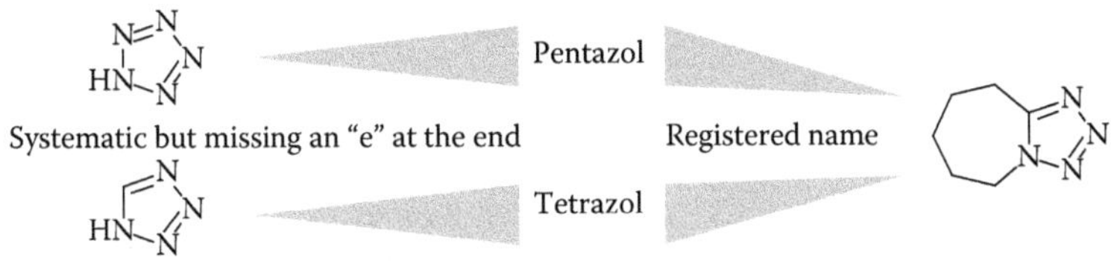

FIGURE 3.10 Alternative treatments of registered names.

are too flexible in their implementation, for example, when the name is not present in a lookup dictionary or ambiguity is not reported, then such labels can be erroneously interpreted as systematic.

Although the two names shown in Figure 3.10 are incorrect according to English IUPAC guidelines for the two structures on the left, they are only *almost* systematic. In fact, in German-language nomenclature where the terminal "e" is not cited, they are correct. However, both of them are listed as registered names for the structure shown on the right and can be found on the ChemIDplus website.[57] Many such examples have spread this problem across the literature and other sources of chemical information. Thus, the support of trivial names is very important even in terms of helping to distinguish real systematic names from false systematic names. However, it would be highly desirable to discontinue the assignment of registered names that mimic systematic names and can therefore be misleading.

SPELLING CORRECTION AND TREATMENT OF PUNCTUATION

In previous sections we examined problems arising as a result of errors in nomenclature. Another significant area is naming errors resulting from misprints or OCR misinterpretation as reviewed earlier in this chapter. Table 3.2 lists the most common naming errors and the reasons for their occurrence.

TABLE 3.2

Typical Name Errors and Their Reasons[a]

Error Type	Main Reason	Example
Missed character	Misprint	**Bn**zene
Character replacement	OCR, misprint	B**c**nzene
Addition of a character	Misprint	Benzene
Inversion of a pair of characters	Misprint	**bn**ezene
Lost space or dash	OCR, misprint	**1**chloropropane
Added space	OCR, misprint	1-chloro propane
Punctuation replacement	OCR, misprint	1**.**2-dichloroethane

[a] The errors in the examples are shown in bold.

Automatic recognition and correction of these errors is a very important component of the chemical name conversion process. Based on available information, this procedure is implemented in the most flexible way in the CambridgeSoft Name=Struct program.[48]

Table 3.3 shows that Name=Struct supports four main types of errors inside chemical names: addition, deletion, replacement, and pair inversion. For the conversion of names generated by OCR, the most common error is character replacement. For example, the name "heptane-2-car**6**oxylic acid" shown in Figure 3.2 and resulting from OCR cannot be converted to a structure.

Other common mistakes are due to the handling of punctuation and enclosing marks. Although their presence is important, the replacement of one type by another generally does not affect the name analysis procedures. The same situation exists with the recognition of enclosing marks where the actual type of enclosing mark has no specific grammatical sense. A well-known exception is that a space is very important for the names of esters, as is shown in the simple example below.

Phenyl acetate   Phenylacetate

The formatting of chemical names is generally not important. Whereas capitalization or italicization are essentially senseless, both sub- and superscripts are helpful in name analysis, and in most cases the absence of formatting can be resolved simply by grammatical implementation. For example, Name=Struct successfully converts polycyclic names like **Tricyclo[3.3.1.11,5]decane** that according to nomenclature rules must be written as **Tricyclo[3.3.1.1^{1,5}]decane**. A good N2S engine therefore

TABLE 3.3
Supported and Unsupported Automatic Error Recognition in Name=Struct

Supported Errors		Unsupported Errors	
Benzioc acid	Pair inversion	Benzoic acdi	Inversion – end or beginning
Benzxoic acid	Letter addition	Benzoic acide	Addition – end or beginning
Benzooic acid	Double letter	Benzoic aci_	Missed – end or beginning
Benzoic a cid	Space	Benzoic acif	Replaced – end or beginning
Bnzoic acid	Missed letter	Bennzoic accid	Two errors in name
Benzoic acld	Replaced letter		

needs to be to be insensitive to both chemical name formatting and punctuation. This can generally be handled very efficiently using name normalization procedures converting all punctuations into one type of separator and all enclosing marks into parentheses.

PROBLEMS ASSOCIATED WITH ASSEMBLING CHEMICAL STRUCTURES

It could be assumed that the conversion of chemical names to their associated structures would conclude the task to provide the necessary data to a chemist to peruse. Unfortunately, the output from N2S engines can be in various formats including SMILES strings, InChI strings, or one of a number of connection table formats. For a chemist to examine a structure, it must be represented in an interpretable graphical format with appropriate spatial configurations including bond angles, bond lengths, *cis/trans* displacements, and stereochemical centers. Although the majority of chemical structure drawing packages integrated with N2S algorithms do include a "cleaning" algorithm, this process is extremely complex, and there is no perfect procedure.[58,59]

One issue that should be noted is the problem of over-determination of a structure, a circumstance that can arise when the generated structure is more specific than the initial chemical name. Part of this problem was described previously in the discussion of the conversion of ambiguous names. A particular problem concerns the assembly of a chemical structure with the appropriate configuration of double bonds. As shown in Figure 3.11, the configuration of the nitrogen–nitrogen double bond is a *trans*-orientation, but the source name did not contain this information. Most N2S engines generate such structures in this situation. In many cases omitted stereoconfigurations in the chemical name mean that either the configuration is unknown or the sample contains a mixture of isomers. The most appropriate result would be to follow the IUPAC guideline for display in the recommended way,[60] but such a depiction is difficult for most procedures used to create "clean" structures. These algorithms remain an area of development for most drawing software development teams.

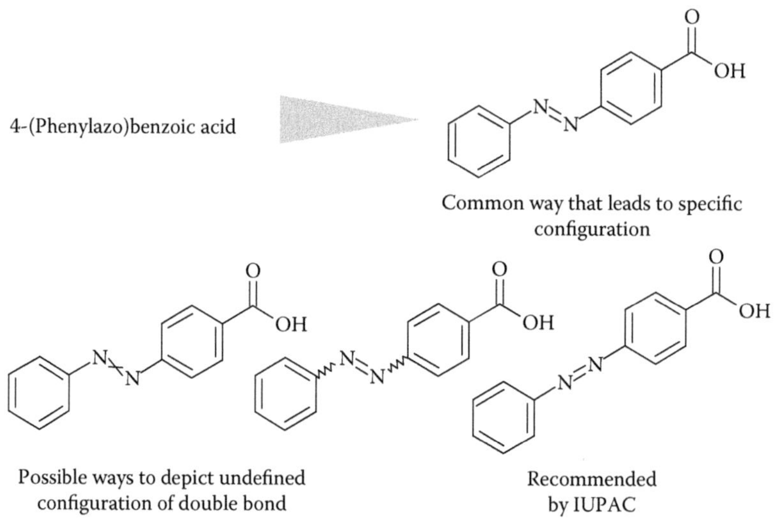

4-(Phenylazo)benzoic acid

Common way that leads to specific
configuration

Possible ways to depict undefined
configuration of double bond

Recommended
by IUPAC

FIGURE 3.11   Graphical representations of undefined versus defined double bond configurations.

CONCLUSIONS

In the near future we can be hopeful that the need to convert chemical names to chemical structures will be less important than we find at present. The ability to encapsulate the majority of organic molecules into an internationally accepted string representing a chemical structure already exists (see Chapter 5 regarding the InChI identifier), and publishers are starting to embed the InChI string directly into their articles to facilitate structure-based communication.[61] Software tools from a number of the commercial vendors can already search across chemical structures embedded in electronic documents and generate either PDF files[62] or image files with chemical structure information embedded directly into those files.[63] As the InChI identifier is extended to include other chemical structures of interest to the community (for example, polymers, organometallics, inorganics, and Markush), the opportunity to further structure-enable all electronic documents for searching is facilitated. As publishers initiate the inclusion of structure-based tags associated with either chemical names or chemical structure depictions, the future of data-mining will require the coordinated extraction of information from documents containing chemical entities in both textual and graphical formats.

Until that time there remains a real need to continue the efforts to convert chemical identifiers, be they names or registry numbers, to their source chemical structures. As optical recognition performance improves, and supporting technologies such as RECAPTCHA[64] contribute to the challenge of text digitization, the conversion of chemical names will be limited only by the quality of the conversion algorithms and the appropriateness of the chemical names. The available N2S conversion algorithms have already demonstrated value and are maturing in capability. The choice of accuracy versus throughput is one for the user. What these algorithms cannot resolve, however, is the potential errors and ambiguities inherent to chemical names present in various documents, and it is the authors' opinion that moving forward, future issues of this nature can only be resolved by adoption of structure identifier embedding inside the document (the suggested format being the InChI identifier), the unlikely development of improved nomenclature skills of all publishing chemists, or, preferably, the adoption of electronic tools for the generation of high-quality systematic names.

While NTS algorithms and other structure mining tools continue to improve, there will likely be many opportunities for errors. Trusting the conversion of chemical names to a computer program without prior knowledge of the nature and quality of the input could be a recipe for disaster when handling publications and, based on our experience, especially when dealing with patents. N2S software is a very useful support aid at best, but quality and validation remain the responsibility of the users of the software, who are responsible for the generation of chemical information via application of the software.

NOTES AND REFERENCES

1. IUPAC names are systematic names generated according to guidelines issued by the International Union of Pure and Applied Chemistry– (http://www.iupac.org). An overview of their nomenclature efforts is provided online at http://en.wikipedia.org/wiki/IUPAC_nomenclature_of_organic_chemistry (accessed December 10, 2007).
2. CA index names are chemicals names issued according to the nomenclature standards of the Chemical Abstracts Service (http://www.cas.org).
3. CAS Registry Numbers are unique numerical identifiers for chemical compounds, polymers, biological sequences, mixtures, and alloys. (http://www.cas.org/newsevents/10digitrn.html; accessed December 11, 2007).
4. An EC-number is a seven-digit code allocated by the Commission of the European Communities for commercially available chemical substances within the European Union. This EC-number replaces the previous EINECS and ELINCS numbers issued by the same organization. (http://ecb.jrc.it/data-collection/; accessed December 11, 2007).
5. The International Union of Pure and Applied Chemistry: http://www.iupac.org/dhtml_home.html (accessed December 11, 2007).
6. Recommendations on Organic & Biochemical Nomenclature, Symbols & Terminology (http://www.chem.qmul.ac.uk/iupac/); The IUPAC Nomenclature Books Series(http://www.iupac.org/publications/books/seriestitles/nomenclature.html).
7. An overview of the history of OCR –can be found at http://en.wikipedia.org/wiki/Optical_character_recognition (accessed December 8, 2007).
8. *Proceedings of the First International Workshop on Text Mining in Bioinformatics (TMBio)*, Arlington, VA. *BMC Bioinformatics,* 2007, 8. Available at http://www.biomedcentral.com/1471-2105/8?issue=S9 (accessed December 25, 2007).
9. ACD/Name, Advanced Chemistry Development, Toronto, Ontario, Canada.
10. The CAS Registry Number and Substance Counts, updated daily, is available at http://www.cas.org/cgi-bin/cas/regreport.pl (accessed December 24, 2007).
11. The Beilstein structure database, Beilstein GmbH, Germany.
12. DiscoveryGate, Symyx, California.
13. The Spresi Database, Infochem GmbH, Germany.
14. VINITI (Vserossiiskii Institut Nauchno-Tekhnicheskoi Informatsii) (All-Russia Scientific Research Institute of Information), Moscow.

15. Elsevier-MDL Patent Chemistry Database, Elsevier, Reed-Elsevier, Amsterdam, The Netherlands; http://www.mdli.com/products/knowledge/patent_db/index.jsp (accessed December 26, 2007).

16. Derwent World Patents Index, Thomson Scientific, The Thomson Corporation; http://scientific.thomson.com/products/dwpi/ (accessed December 23, 2007).

17. The SureChem Portal, SureChem Inc., San Francisco, CA; http://www.surechem.org (accessed December 10, 2007).

18. IBM Chemical Search Alpha, IBM, Almaden Services Research, San Jose, CA; https://chemsearch.almaden.ibm.com/chemsearch/SearchServlet.

19. Goncharoff, N. 2007. SureChem—free access to current, comprehensive chemical patent searching, presentation given at the ICIC meeting, Infonortics, Barcelona, Spain.

20. Boyer, S. 2007. IBM, Almaden Services Research, San Jose, CA, presentation given at the ICIC meeting, Infonortics, Barcelona, Spain. Available at http://www.infonortics.com/chemical/ch04/slides/boyer.pdf (accessed December 3, 2007).

21. Advanced Chemistry Development, Toronto, Ontario, Canada; http://www.acdlabs.com.

22. CambridgeSoft, Boston, MA; http://www.cambridgesoft.com.

23. Openeye, Santa Fe, NM; http://www.eyesopen.com.

24. World Intellectual Property Organization (WIPO); http://www.wipo.int/about-wipo/en/what_is_wipo.html (accessed December 29, 2007).

25. Medline/Pubmed; http://www.ncbi.nlm.nih.gov/sites/entrez.

26. SureChem, Nicko Goncharoff, private communication (December 2007).

27. Rhodes, J., Boyer, S., Kreulen, J., Chen, Y., and Ordonez, P. 2007. Mining patents using molecular similarity search. In *Proceedings of the 12th Pacific Symposium on Biocomputing*, 12:304–315.

28. An overview page regarding the IBM Chemical Search Alpha is available at https://chemsearch.almaden.ibm.com/chemsearch/about.jsp.

29. Accelrys, San Diego, CA; http://www.accelrys.com (accessed December 16, 2007).

30. Accelrys' Scitegic Pipeline Pilot ChemMining Collection; http://www.scitegic.com/products/chemmining/ChemMining.pdf (accessed December 15, 2007).

31. Murray-Rust, P., Mitchell, J. B., and Rzepa, H.S. 2005. Communication and re-use of chemical information in bioscience. *BMC Bioinformatics,* 6:180.

32. Corbett, P. and Murray-Rust, P. 2006. High-throughput identification of chemistry in life science texts, in *Computational Life Sciences II*, Lecture Notes in Computer Science series. Berlin/Heidelberg: Springer, pp. 107—118.

33. PubChem: information on biological activities of small molecules; http://pubchem.ncbi.nlm.nih.gov/ (accessed December 12, 2007).

34. OPSIN, an Open Parser for Systematic Identification of Nomenclature; http://depthfirst.com/articles/2006/10/17/from-iupac-nomenclature-to-2-d-structures-with-opsin (accessed December 10, 2007).

35. MDL was acquired by Symyx in 2007; http://www.symyx.com/press_release.php?id=4&p=255 (accessed December 12, .2007)

36. TEMIS S.A. Skill Cartridge Chemical Entity Relationships; http://www.temis.com/fichiers/t_downloads/file_109_Fact_sheet_TEMIS_Skill_Cartridge_Chemical_Entity_Relationships_En.pdf (accessed December 16, 2007).

37. The National Center for Text Mining, United Kingdom; http://www.nactem.ac.uk/software.php?software=namedentity (accessed December 16, 2007).

38. Optical character recognition; http://en.wikipedia.org/wiki/Optical_character_recognition (accessed December 10, 2007).

39. Gifford-Fenton, E. and Duggan, H. N. Electronic textual editing: effective methods of producing machine-readable text from manuscript and print sources. Available at http://www.tei-c.org/About/Archive_new/ETE/Preview/duggan.xml.

40. OmniPage SE Version 2.0 by ScanSoft Inc. (ScanSoft Inc. has since merged with Nuance and assumed their name); http://www.nuance.com/company/) (accessed December 29, 2007).

41. Hodge, G., Nelson, T., and Vleduts-Stokolov, N. 1989. Automatic recognition of chemical names in natural language text. Paper presented at the 198th American Chemical Society National Meeting, Dallas, TX, April 7–9.

42. Kemp, N. and Lynch, M. F. 1994. The extraction of information from the text of chemical patents. 1. Identification of specific chemical names. *Journal of Chemical Information and Computer Sciences*, 38(4):544–551.

43. Corbett, P., Batchelor, C., and Teufel, S. 2007. Annotation of named chemical entities, in *BioNLP 2007: Biological, Translational, and Clinical Language Processing*, pp. 57–64, Prague, Czech Republic.

44. Copestake, A., Corbett, P., Murray-Rust, P., Rupp, C. J., Siddharthan, A., Teufel, S., and Waldron, B. 2006. An architecture for language processing for scientific texts. *Proceedings of the UK e-Science Programme All Hands Meeting*, Nottingham, UK.

45. Garfield, E. 1961. Chemico-linguistics: computer translation of chemical nomenclature. *Nature*, 192:192–196.

46. Vander Stouw, G. G., Naznitsky, I., and Rush, J. E., 1967. Procedures for converting systematic names of organic compounds into atom-bond connection tables. *Journal of Chemical Documentation,* 7(3):165–169.

47. Vander Stouw, G. G., Elliott, P. M., and Isenberg, A. C. 1974. Automated conversion of chemical substance names into atom-bond connection tables. *Journal of Chemical Documentation,* 14(3):185–193.

48. Structure=Name Pro version 11.0, CambridgeSoft, Boston, MS; http://www.cambridge soft.com/software/details/?ds=0&dsv=122 (accessed December 12, 2007).

49. Brecher J. S. 2006. Method, system, and software for deriving chemical structural information. United States Patent 7054754, issued May 30.

50. ACD/Name (Advanced Chemistry Development). Generate Structure From Name; http://www.acdlabs.com/products/name_lab/rename/tech.html (accessed December 16, 2007).

51. NamExpert, ChemInnovation Software, San Diego, CA; http://www.cheminnovation.com/products/nameexpert.asp (accessed December 16, 2007).

52. Bonniot, D. IUPAC naming. Presented at the ChemAxon UGM Meeting, Budapest, Hungary. Available at http://www.chemaxon.com/forum/viewpost12244.html#12244 (accessed December 16, 2007).

53. IC$_{N2S}$, InfoChem GmbH, Munchen, Germany; http://infochem.de/en/mining/icn2s.shtml (accessed December 14, 2007).

54. Brecher, J. S. 1999. Name=Struct: a practical approach to the sorry state of real-life chemical nomenclature. *Journal of Chemical Information and Computer Sciences (JCICS)*, 39:943–950.

55. Eller, G. A. 2006. Improving the quality of published chemical names with nomenclature software. *Molecules*, 11:915–928.

56. Thielemann, W. 2007. Information extraction from full-text. Presented at *The International Conference in Trends for Scientific Information Professionals*, Barcelona, Spain. Available at http://www.infonortics.com/chemical/ch07/slides/thielemann.pdf (accessed December 6, 2007).

57. ChemIDplus, National Library of Medicine, Maryland; http://chem.sis.nlm.nih.gov/chemidplus/ (accessed December 16, 2007).

58. Clarck, A. M, Labute, P., and Sabtavy, M. 2006. 2D structure depiction. *Journal of Chemical Information and Computer Sciences,* 46:1107–1123.

59. Fricker, P. C, Gastreich, M., and Rarey, M. 2004. Automated drawing of structural molecular formulas under constraints. *Journal of Chemical Information and Computer Sciences,* 44:1065–1078.

60. Brecher, J. S. Graphical representation standards for chemical structure diagrams (IUPAC Recommendations 2007). Provisional recommendations. Available at: http://www.iupac.org/reports/provisional/abstract07/brecher_300607.html (accessed December 25, 2007).

61. Project Prospect, Royal Society of Chemistry, Cambridge, UK; http://www.rsc.org/Publishing/Journals/ProjectProspect/Features.asp (accessed December 10, 2007).

62. ACD/ChemSketch, Advanced Chemistry Development, Toronto, Ontario, Canada; http://www.acdlabs.com/products/chem_dsn_lab/chemsketch/features.html#Reporting (accessed December 10, 2007).

63. Wikipedia Discussions Forum. 2007. Embedded InChIs in images; http://en.wikipedia.org/wiki/Wikipedia_talk:WikiProject_Chemistry/Structure_drawing_workgroup/Archive_Jun_2007#ACD_ChemSketch_-_the_company_is_willing_to_make_a_Wikipedia_Template_on_their_Freeware.

64. RECAPTCHA™. Digitizing Books One Word at a Time (accessed December 16, 2007).

4 Identification of Chemical Images and Conversion to Structure-Searchable Information

A. Peter Johnson and Anikó T. Valkó

CONTENTS

INTRODUCTION

Depictions of two-dimensional chemical structures published in the literature are stored as bitmap images in nearly all electronic sources of chemical information such as reports, journals, and patents. Although chemical structures for publication are usually created using chemical drawing programs that generate complete structural information, this information is lost in the publication process. The published structures are normally in the form of bitmap images that are easily interpreted by humans but lack the explicit structural information required for input to chemical analysis software packages or chemical databases. The reproduction of this information by redrawing the structure with a computer program is time-consuming and prone to errors but nevertheless is still the norm for these purposes.

Clearly, there is a pressing need for an equivalent to optical character recognition, optical chemical structure recognition, that can automatically turn bitmapped structural diagrams into structure descriptions—connection tables or equivalent structural strings—that are suitable for input into chemical structure databases.

PROJECTS

Interest in optical structure recognition dates back to the early 1990s when four projects were developed and published: the Contreras system, by M.L. Contreras et al. (1990), Kekulé, by J.R. McDaniel and J.R. Balmuth (McDaniel and Balmuth 1992; Borman 1992), the IBM system, by S. Boyer et al. (Casey et al. 1993), and CLiDE (Chemical Literature Data Extraction), by A.P. Johnson et al. (Ibison et al. 1993, 1992; Kam et al. 1992). The Contreras, Kekulé, and IBM systems were mainly aimed at extracting information from chemical structures in the chemical literature. The CLiDE system was more ambitious in that; the long term goal was to process whole pages of chemical information from journals and books. This required the ability to deal not only with chemical structures in the literature, but also with reaction schemes and other relevant chemical information in the text.

A project called chemoCR (Algorri et al. 2007a, b; Zimmermann et al. 2005) was started in 2004 at the Algorithms and Scientific Computing Institute of the Fraunhofer Society. The validation of chemoCR on a large set of chemical images shows encouraging results. However, the system in its current stage has apparently only been tested on images directly produced by drawing programs, which means its performance on scanned documents, with all the distortion and restoration problems they include, has yet to be demonstrated. Also, the system does not accept as input complete document pages containing text, tables, and images, but rather just the images of chemical molecules on their own.

A new version of CLiDE, CLiDE Pro, was initiated in the beginning of 2006 by the software company Keymodule Ltd. The primary aim of the CLiDE Pro project is to overcome some of the limitations of CLiDE, and thus achieve a high recognition performance on a diverse set of structure diagrams, and to include additional recognition capabilities, for instance, in the area of patents. This project is still in progress, and no results have been reported in the literature as yet.

Recently, two free open source programs, ChemReader and OSRA, were released. ChemReader[*], developed at the University of Michigan, is based on a machine vision approach using empirically derived chemical intelligence to identify image features that are both accurately resolvable and highly informative for indexing a specific chemical database such as NCBI's PubChem[†] database. Optical Structure Recognition Analysis (OSRA)[‡] is the latest addition to the optical structure recognition tools. OSRA can read a document in any one of the over 90 graphical formats that can be parsed by ImageMagick,[§] including GIF, JPEG, PNG, TIFF, PDF, and PS, and generate the Simplified Molecular Line Entry System (SMILES) representation of the molecular structure images encountered within that document. Tests on OSRA have been conducted by Antony Williams, and the results are reported on the ChemSpider blog[¶]. Although OSRA appears to work reasonably well on clearly drawn structure diagrams, Williams identifies problems with stereo bonds, crossing bonds, and metallo-organic structures. In conclusion, OSRA at its current stage is rated as a "work in progress," but the general approach and the idea of a tool freely available to the chemistry community is appreciated widely in chemistry blogs and scientific forums.

PROBLEM OVERVIEW

Before optical structure recognition methods can be applied to a document, the chemical images in the document must be identified and separated from the rest of the document so subsequent processing stages can operate exclusively on the

---

[*] Michigan Alliance for Chemoinformatic Exploration (MACE), University of Michigan (http://www. stat.lsa.umich.edu/~kshedden/MACE/).

[†] Information on biological activities of small molecules. National Center for Biotechnology Information (NCBI) (http://pubchem.ncbi.nlm.nih.gov/).

[‡] SAIC-Frederick, NCI-Frederick, NIH, DHHS (http://cactus.nci.nih.gov/osra/).

[§] Image manipulation library. ImageMagick Studio LLC (http://www.imagemagick.com/).

[¶] Database of Chemical Structures and Property Predictions. ChemZoo Corporation (http://www.chem spider.com/).

chemical graphic information. The core problem in the field of optical chemical structure recognition is the compilation of chemical graphs of individual molecules from chemical images. The retrieved chemical graphs are then used to interpret complex objects such as generic structures and reaction schemes, in conjunction with information related to the complex objects (e.g., reaction arrows in reactions, descriptions of substituents of R-groups of generic structures). Finally, extraction of bibliographic information (image caption, author, document title, abstract, journal name, page number, etc.) by logical document layout analysis (Simon and Johnson 1997; Simon 1996) is desirable for automatic document handling, automatic retrieval of bibliographic information, and support for hierarchical browsing.

INDIVIDUAL STRUCTURES

A structure diagram conveys the exact structural nature of a particular chemical compound through a drawing. Although the conventions for this type of description are not clearly defined (Loening 1988), chemists are able to correctly interpret the wide variety of drawing styles commonly found in the chemical literature. The information contained in a structural diagram of a compound can be divided into three areas: atom information, bond information, and structural information. All three types of information have to be retrieved to extract a molecule from a structure drawing.

Atom information includes the chemical element's name (e.g., N) of individual atoms as well as functional groups denoted by character strings (e.g., MeO) and representations of generic groups (e.g., R). Printed material also frequently contains ancillary information such as the vertex label of an atom. In addition, labels denoting features such as atomic weight, charge, chirality, hybridization, and valency are sometimes part of the atom information in the image. Figure 4.1 shows some examples of structure diagrams that contain different types of atom information.

Bond information includes the bond order, that is, single, double, or triple bond, and bond style, such as simple straight-line bond, wedged bond, dashed bond, wavy bond, broken-line bond, bold bond, and so on. Bond information can also include some special bond types representing the aromatic bonds or bond stereochemistry, for example, wedged or dashed bond. Bond labels are also sometimes found in structures. Different types of bond information are illustrated in Figure 4.2.

Atoms and bonds connected to each other form a structure. Therefore, atom information and bond information are also part of the structure information. Apart from this, a structure can have additional information that relates to the whole structure, such as overall charge or structure label (see Figure 4.3). The latter is routinely used as a means of referencing a structure in the main body of the text.

GENERIC STRUCTURES

A generic substance represents more than one substance or a set of specific substances. This set of substances can be represented by a generic or Markush structure. An important use of generic structures in the literature is that they provide a compact representation of a set of specific substances. They are also commonly used to show the way the value of a particular physical or biological property varies as a particular

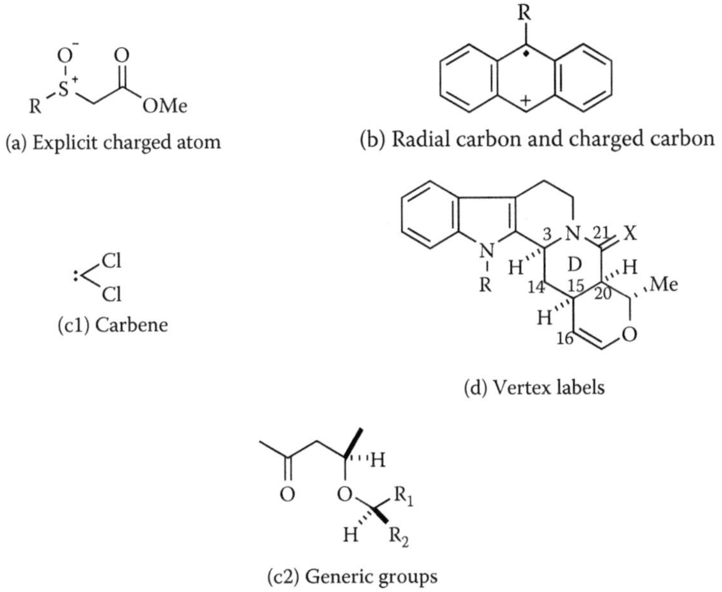

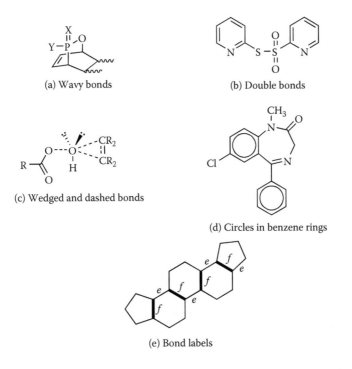

(a) Explicit charged atom

(b) Radial carbon and charged carbon

(c1) Carbene

(d) Vertex labels

(c2) Generic groups

FIGURE 4.1　Structure diagrams containing different types of atom information.

(a) Wavy bonds

(b) Double bonds

(c) Wedged and dashed bonds

(d) Circles in benzene rings

(e) Bond labels

FIGURE 4.2　Structure diagrams containing different types of bond information.

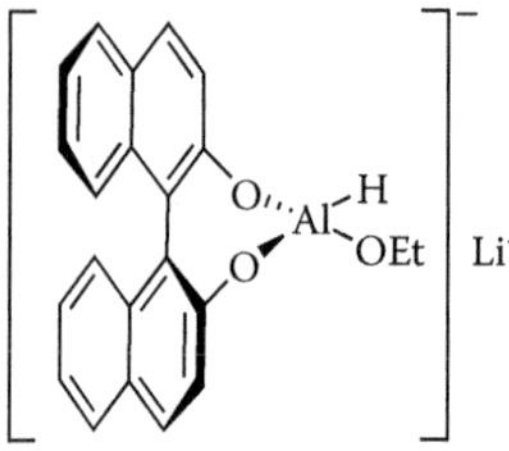

(a) Overall charge

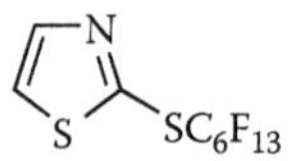

(b) Structure label

FIGURE 4.3   Structure information.

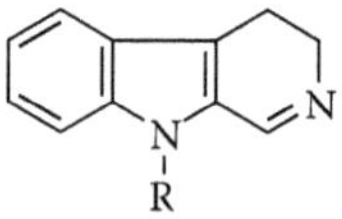

22: R = H
33b: R = CO$_2$Me
33c: R = CH$_2$OMe

FIGURE 4.4   A typical generic structure.

substituent is varied. The large body of research carried out in this area includes methods for the storage and retrieval of generic chemical substances (Fisanick 1990; Barnard et al. 1982), methods to automatically interpret the text concerning the generic structures of chemical patent abstracts (Chowdhury and Lynch 1992a,b), and the design of a formal language, GENSAL, to provide a concise and unambiguous representation of generic structures from chemical patents (Lynch et al. 1981; Barnard et al. 1981).

A generic structure generally is composed of an invariant part together with associated variable groups, or R-groups. The R-groups indicate the possible alternatives for the invariant part. In most cases, the structure lies inside a graphic region, and the R-groups lie inside a text region (see Figure 4.4). The R-groups attached to the invariant part are marked with symbols, and their substitution values are given separately.

Markush structures from different sources tend to use different mechanisms for introducing variability. Combinatorial libraries are generally the simplest, having only *substituent variation* with a fixed list of specific alternatives. Patents are the most complex, involving all types of variation. *Position variation* can be introduced by variation of attachment position in the core structure. *Frequency variation* can be used to vary the multiplicity of occurrence of a group. *Homology variation* involves the use of generic expressions, such as *alkyl* or *aryl*, which represent a potentially unlimited class of radicals, characterized by common structural features. Homology variation is the most difficult to deal with, because a single expression can encompass a potentially infinite set of specific alternatives, rendering enumeration-based approaches unfeasible. Figure 4.5 shows a generic structure containing all variation types.

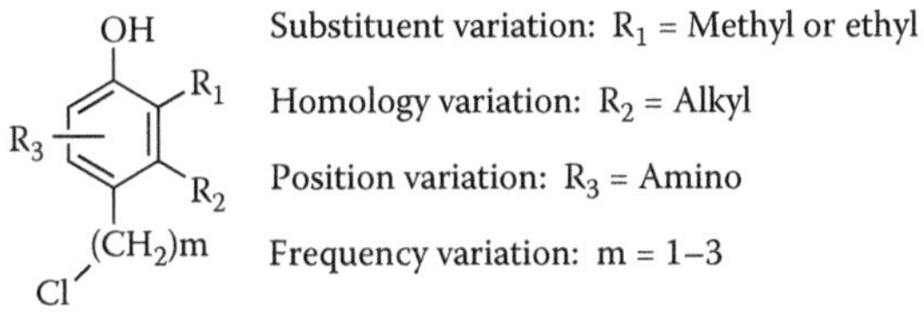
OH
R_1
R_3
R_2
$(CH_2)m$
Cl
Substituent variation: R_1 = Methyl or ethyl
Homology variation: R_2 = Alkyl
Position variation: R_3 = Amino
Frequency variation: m = 1–3

FIGURE 4.5 A Markush structure illustrating the four types of variability of R-groups.

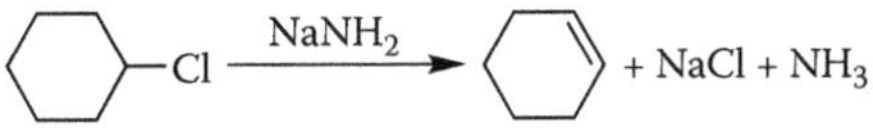
Cl
$NaNH_2$
+ NaCl + NH_3

FIGURE 4.6 A chemical reaction with three products.

TABLE 4.1
Arrow Types Most Frequently Used in Reactions

Simple arrow	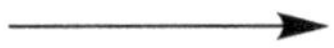	
Bent arrow	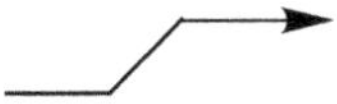	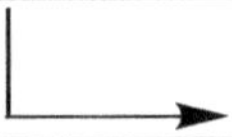
Failed reaction arrow	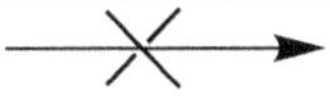	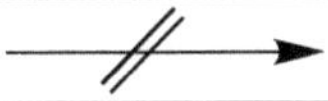
Equilibrium arrow	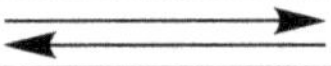	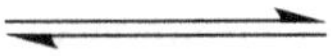
Crossed equilibrium arrow	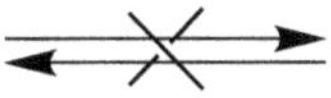	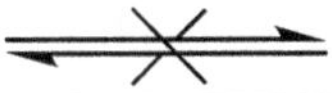
Complex arrow	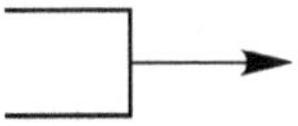	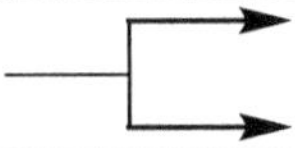
Resonance arrow	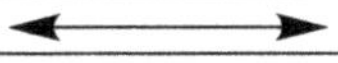	
Multistep arrow	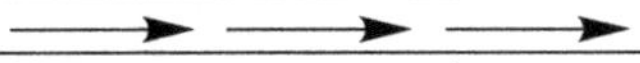	
Retrosynthetic arrow	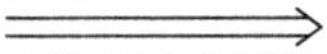	
Dashed arrow	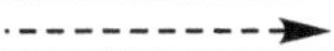	

Reaction Schemes

The key components of a chemical reaction are the reactants and products, and these are usually represented by their chemical structures. In some cases, the products or reactants might be represented by a chemical name or a number indicated in the related text. When there are several reactants or products in a reaction, those structures within such a group are usually joined by the symbol "+" (see Figure 4.6), which is called the "joiner."

An arrow in a reaction is used as a separation symbol between the product and the reactant. Structures situated in front of the arrow head are products, and those at the rear of the arrow tail are reactants. An arrow's presence indicates the existence of a reaction. Arrows can have different types (see Table 4.1). The most common one, which contains one head and one tail only, is called a *simple arrow*. An arrow is called a *bent arrow* if its head and its tail are not on one straight line. The type of arrow might also convey information, such as whether it is a multistep reaction, an unsuccessful reaction, or a retrosynthetic reaction. It might additionally indicate approximately how many reactions are represented by the arrow. A simple arrow represents just one reaction, whereas a *complex arrow* containing more than one tail or one head may represent more than one reaction.

In many reactions, reagents and conditions are indicated along reaction arrows. The reagents used in a reaction may be represented by their chemical names or chemical structures as shown in Figure 4.7.

The information required to fully characterize a reaction includes a list of connection tables of the reactants, a list of connection tables of the products, the arrow type for the reaction, the reagent text, and any graphic reagents for the reaction. Other useful information might include reaction time, temperature, yield of product, quantities of reactants, apparatus, energy, atmosphere, and catalyst. This information is usually written in the text, and the extraction of this kind of information involves interpretation of the text.

Reaction schemes are composed of several reactions. Once the information concerning individual reactions is available, it should be simple to generate information for the complete scheme. Figure 4.8 shows a reaction scheme consisting of three reactions.

INPUT DATA

The most common input data to the systems introduced in "Projects," above, is the two-dimensional digital raster images of chemical molecules and whole journal pages. Systems are often limited to handle bilevel images containing *on*-pixels (black) and *off*-pixels (white) because molecule diagrams are usually drawn with black ink on a white background. However, colors might be used in structure drawings for various purposes, for example, for indicating atom types or for highlighting important structure parts. Furthermore, it is also possible for a color other than white to be used as a background color. To distinguish the individual elements of a molecule in a color-scaled image from the background, the image has to be binarized. During *binarization* (Tetsuo et al. 1996), a color-scaled image is turned into a bilevel image by classifying every pixel as an *on*-pixel or as an *off*-pixel. Two systems, CLiDE Pro and chemoCR, can handle color-scaled images, applying a threshold-based binarization technique and an adaptive histogram binarization algorithm, respectively.

Because the vast majority of published articles are available for download in PDF (portable document format) file format, the ability to feed PDF files into a

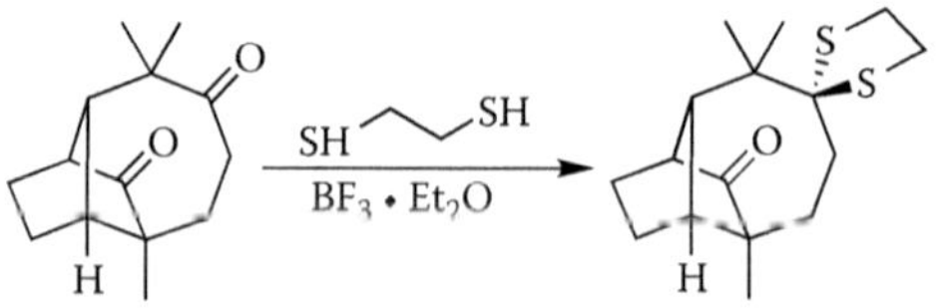

FIGURE 4.7   Reaction with reagent text and graphic reagent.

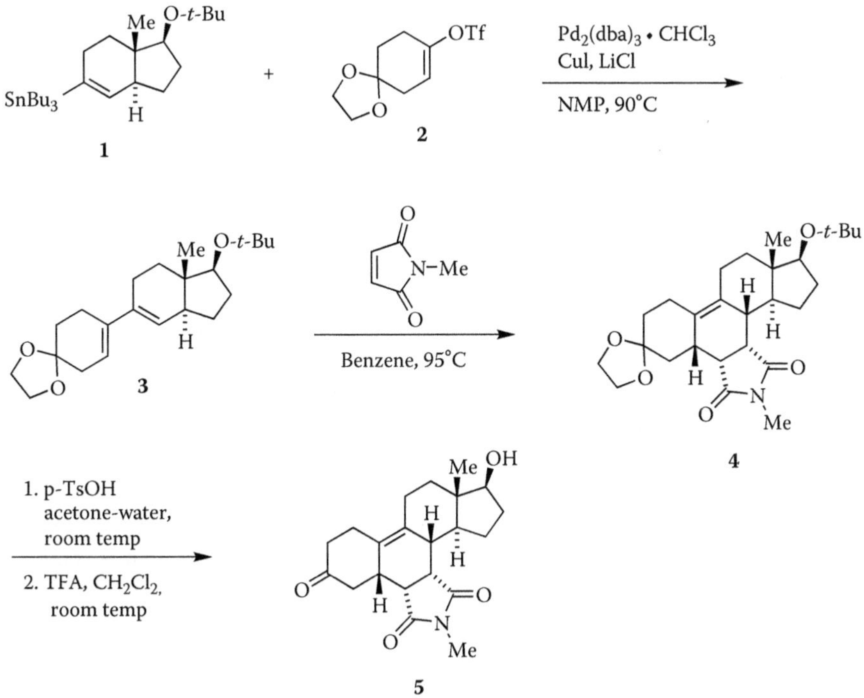

FIGURE 4.8 A reaction scheme.

structure-recognition program is highly desirable. The PDF documents of old articles consist entirely of scanned images of text and graphics. However, recent PDF documents contain textual information, tables, and images separately and distinguish images from the rest of a journal page, which permits the extraction of text information directly, thus avoiding the need for optical character recognition (OCR) on the text or document image segmentation (see the following section).

IDENTIFICATION OF CHEMICAL IMAGES

To identify chemical diagrams, a digitized document page consisting of a mixture of text and graphics has to be segmented to apply the appropriate recognition technique to each part. *Document image segmentation* methods published in the literature can be divided into top-down and bottom-up techniques. The *top-down* methods look for global information on the page such as black and white stripes, and on the basis of this split the page into blocks that are successively divided into subblocks, in an iterative fashion, to obtain the final text and graphics segments (Ittner and Baird 1993; Baird 1994; Krishnamoorthy et al. 1993; Nagy et al. 1992; Pavlidis 1968). The *bottom-up* approach relies on a data-driven technique that refines the data by layered grouping operations. In practice, single pixels are gathered on the basis of a low-level analysis, to constitute blocks that can be merged into successively larger blocks (Fan et al. 1994; Fletcher and Kasturi 1988; O'Gorman and Kasturi 1993; Saitoh et al. 1994; Tsujimoto and Asada 1992).

From the published data, it appears that only two of the systems, CLiDE and the IBM system, introduced in "Projects," above, perform document image segmentation to automatically identify graphic regions in document pages.

The document segmentation method implemented in CLiDE (Simon 1996; Simon et al. 1995) builds up the tree structure of a page in a bottom-up manner; that is, it starts by processing the connected on-pixel regions or *connected components* of the image, and results successively in a list of words, text lines, text and graphic blocks, and columns. Once the image is loaded, the connected components of the page are found and the noise-like connected components are removed. The layout analysis starts with the calculation of the distances between the pairs of connected components using the enclosing boxes around the connected components. If the connected components are considered as the vertices of the graph, and the distances between them as the weighted edges of the graph, then words, lines, blocks, and so on can be derived from the minimal-cost spanning tree, built with Kruskal's algorithm (Aho et al. 1983).

In the IBM system (Casey et al. 1993), the scan array is resolved into connected components that are defined as many-sided convex polygons bounding the enclosed subimage. These *bounding polygons* are characterized by bands, that is, pairs of opposite parallel sides with each pair at one of a fixed set of directions. Each polygon is the intersection of several such bands that are approximately equally spaced. The system searches for a connected component whose maximum dimension exceeds a threshold d. The parameter d is chosen to exceed the maximum character size expected on the page. Consequently, a subimage satisfying the threshold test can be assumed to be a section of a chemical structure. A search is then made for neighboring connected components within a specified distance threshold from the selected component. The distance threshold is also a parameter of the system, chosen to be smaller than the white space that separates diagram elements from surrounding text. Any components satisfying this test are combined with the initial connected component to define an enlarged bounding polygon that contains the entire group. The search then iterates using the expanded region. This region-growing process terminates when no further connected components are found within the margin determined by the distance threshold.

Document image segmentation merely discriminates the contents of graphical regions, without answering the question of whether a graphic region contains a chemical diagram. Gkoutos et al. (2003) report an approach that uses computer vision methods for the identification of chemical composition diagrams from two-dimensional digital raster images. The method is based on the use of Gabor wavelets (Jain and Bhattacharjee 1992) and an energy function to derive feature vectors from digital images. These are used for training and classification purposes using a Kohonen network for classification with the Euclidean distance norm.

KEY STEPS IN THE RECOGNITION OF INDIVIDUAL STRUCTURES

As described in "Individual Structures," above, the recognition of a molecule from a chemical drawing requires the extraction of three kinds of information: the atom information, the bond information, and the structure information. Generally, this involves the following four steps:

- Classification of components
- Graphic recognition, including vectorization and dashed-line construction
- OCR
- Compilation of connection tables or chemical graphs

During *component classification*, the connected on-pixel regions, or connected components, drawn in the image to represent a molecule in two dimensions are analyzed to determine that connected components are characters, which are lines or graphical shapes, and which are noise. The classification means that regions in the structure diagram illustrating atom information using a string of characters are distinguished from regions containing bond information with lines and graphical shapes. Also, regions containing noise are separated and ignored in later stages.

During *graphic recognition*, all the basic elements of bond information, the elementary lines and curves, as well as dashed lines and wedges are extracted.

The process of decomposing line-drawing images into primitive graphic elements such as lines and curves is often called *image segmentation,* or *vectorization.* Most methods for vectorization start by reducing the width of the line-like connected components from many pixels to just a single pixel. This process is called *thinning* (Naccache and Shinghal 1984; Smith 1987) or *skeletonization.* The skeleton of the image is then segmented into straight lines and curves by finding the dominant points. This method may not be the best choice for chemical diagrams because thinning algorithms are time-consuming, and during the process important information is lost about the diagram; for example, wedged and wavy bonds are important in chemical structures, and this information is not present in the skeleton of a drawing.

Dashed-line construction is performed on connected components recognized to have small dimensions or already classified as dashes to convert them to single picture elements instead of unconnected small lines. Several techniques are available for finding elements of collinear lines, including Hough transforms (Duda and Hart 1972; Illingworth and Kitter 1988). The theory behind the Hough transform is that points on a line, transformed from XY into r-θ space, will result in peaks that can be distinguished from noncollinear data.

OCR (Govindan and Shivaprasad 1990) is performed on connected components previously classified as characters. At this stage, individual characters are assembled into character strings based on XY coordinates; that is, the XY positions of various individual characters are compared, and character strings are assembled based primarily on adjacency of the coordinates.

The function of *connection table building* is to correctly identify the chemical context of the texts and graphics included in a chemical drawing and to create a connection table or chemical graph from them.

The order of the four recognition steps and the set of connected components on which the steps are operated is not the same in all systems. For example, the first step of Kekulé (see "Kekulé," below) is the vectorization performed on all of the connected components. Partial connected component classification is performed during dashed-line construction to identify dashed-line connected components, and during OCR to identify characters. Connected components other than dashed lines and characters are treated as a set of vectors and subjected to the connection table building step.

The following sections describe the structure recognition methods implemented in the various systems.

Kekulé

The process of interpreting a chemical structure diagram in Kekulé (McDaniel and Balmuth 1992; Borman 1992) consists of four steps:

- Vectorization
- Dashed-line construction
- OCR
- Graph compilation

Vectorization

Vectorization reduces the scanned image to line elements only 1 pixel in width by thinning and then forming the straight line segments, called *vectors*, for the image. The results of this step is lists of vectors associated with the original pictorial elements by coordinates of the vector end points. An adaptive smoothing algorithm developed by the authors eliminates the fine, pixel-level detail inherent in a direct translation from pixels to vectors.

Dashed-Line Construction

Dashed-line construction is composed of an exhaustive search over the subset of features that might be possible constituents of a dashed-line or dashed-wedge feature. In general, all dashes that consist of at least two line segments are identified. This includes most cases where one of two line segments is attached to other bonds.

OCR

The characters are first normalized by rotating the original scanned image to correct for scanning error and by combinations of scaling under sampling and contrast and density adjustments of the scanned characters. In operation, the normalized characters are then presented to a multilayer perceptron neural network for recognition; the network was trained on exemplars of characters form numerous serif and sans serif fonts to achieve font invariance. Where the output from the neural network indicates more than one option, for example "5" and "s," the correct interpretation is determined from context.

Individual characters are assembled into character strings based on the adjacency of the coordinates. The implementation can also handle subscripts and superscripts.

Graph Compilation

Graph compilation is the process of interpreting the remaining vector data—after eliminating vectors associated with characters or character strings identified in the preceding step—into a connection table. It is divided into five steps:

1. Each character string resulting from the OCR processing is defined as a node.
2. The remaining vectors, that is, those not belonging to a character string or dashed line, are assumed to represent bonds. The list of these remaining vectors is examined to determine whether either end of any vector is near a character string–defined node. If it is, then it is attached to that node; if not, a new node is created at the end of the vector.
3. The pixels in the original image of a line indicate the line width. The line width at the end points of a line is used to determine whether it is a wedged bond. This is done when a new connection is made between nodes.
4. When connecting vector ends to nodes, the connection is upgraded from a single bond (initial assumption) to a double bond and, finally, to a triple bond if two or three coincident connections between nodes are found.
5. Dashed lines that have previously been interpreted are added to the graph.

The graph compilation ends with a postprocessing step that uses current graph information and the partially processed character string data, such as atom symbols with a charge or group formula, to determine the character string's chemical meaning. The group formula is interpreted into graph format. The existing graph is then analyzed to look for circles (and convert them to alternating single–double bonds), nodes that are really bond crossings, and large parentheses and brackets.

THE CONTRERAS SYSTEM

The recognition process implemented in the Contreras system (Contreras et al. 1990) is iterative, recognizing a subgraph during each iteration. After all of the regions of the image are processed, the system integrates the several subgraphs into the corresponding structure. The following steps are performed during one iteration:

- Component classification based on contour search
- Vectorization based on vertex determination
- Subgraph recognition
- OCR

Component Classification

A left-to-right horizontal sweep is done on the digitized image. It starts from the upper-left part of the image until it finds the first on-pixel. A counterclockwise contour search algorithm is then applied to record the coordinates of every pixel of the contour until arriving back at the first pixel. If the number of pixels of the contour is large enough, it is interpreted as a graph contour. Otherwise, it is considered a chemical symbol.

Vectorization

In the case of graph contours, vertices are located by searching for deflections of the linear trajectory of the external or internal contours around the objects. Vertices are found during perception, when a deflection angle of the linear trajectory higher than a predefined parameter value is detected. The default deflection angle is 18 degrees.

Trajectory is determined by the pixel's neighborhood connectivity method of image processing, where the direction of displacement is described by a number from 0 to 7. This is in function of the relative position of any of the eight neighbors of a central point. The number chosen for the neighbor that is bonded to the central point describes a line coincident with the direction of the displacement.

External and internal borders of a graph arc are used to detect the type of bond associated with a molecular structure (wedge, dot, single, and multiple) as a function of the thickness of the bond and the number of lines joining the atoms.

Subgraph Recognition

Two or more vertices within a defined small space indicate the location of an atom in the structure (see Figure 4.9c). Atoms are numbered, and the neighborhood relationship among them is kept. Any atom with only a single neighbor is considered a possible terminal atom. A linear projection of its previous bond up to a distance similar to the length of that bond is made (see Figure 4.9d). If no on-pixels are encountered, the atom is considered a default carbon atom in the structure. Otherwise, the contour determination process is done over the newly found on-pixel. In this way, the detected chemical symbols are submitted to the OCR module.

Multiple bonds, internal rings, and other molecular substructures or subgraphs (see Figure 4.10a) are recognized through a circular inspection method. A circle of inspection centered on each detected atom is considered (see Figure 4.10b). Unknown border pixels found in this way are kept and used as the initial point for a new counterclockwise contour search, and the perception of new vertices and probable new atoms is carried out as described earlier.

OCR

The OCR process is divided into two steps: separation of each character into a matrix and the recognition of each matrix. Character separation is performed because characters are often overlapped, and therefore recognition is difficult or unreliable. Each separated matrix undergoes a noise filtration to eliminate isolated on-pixels. The character recognition is achieved by feature extraction of the refined matrix.

THE IBM SYSTEM

The system developed at the IBM Almaden Research Center (Casey et al. 1993) performs the following steps to extract chemical structures from diagrams:

- Vectorization
- Component classification based on vectors
- OCR
- Connection table building

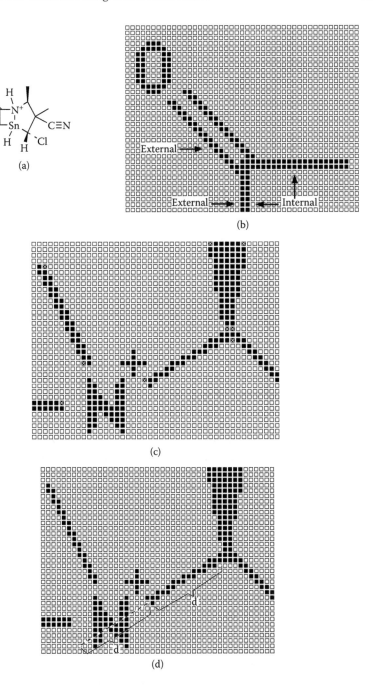

FIGURE 4.9 Vectorization and subgraph recognition in the Contreras system. (a) A molecular structure to be recognized. (b) Partial view of the digitized structure showing the external and internal border of a bond. (c) Digitized partial view of the graph showing vertices position at the x points. (d) Perception of an atom by linear projection and search of on-pixels along the path.

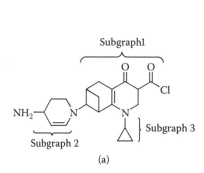

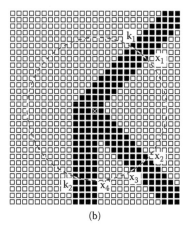

(a)   (b)

FIGURE 4.10   Circular inspection method in the Contreras system. (a) A molecular structure with three subgraphs. (b) Circular sweeping, where k_i and x_i represent known and unknown points, respectively.

Vectorization

Vectorization is done by the Graphical Image Formatting and Translating System (GIFTS), developed by IBM Tokyo Research Laboratory. The GIFTS algorithm fits lines to the given pixel array, producing as output a set of end point coordinates. End points of lines are labeled as *free ends*, *junctions* (where three or more lines meet), *loop closures*, or *connections* (two lines meeting).

Since the GIFT vectorizing algorithm is based on very general principles, vectors identified as character vectors (see classification of components, below) are analyzed, and errors detected in these vectors are corrected to obtain a proper representation of bond structure. This cleanup stage corrects two types of defects:

- Breaking of lines in the region of a junction
- Breaking of a single diagram line into two or more vectors at points away from a junction

The first objective is accomplished by detecting any vector with a length less than a specified fraction of the median value of a diagram line. Such a vector is, in effect, shrunk to a single point, its midpoint. That is, the terminal of any vector connected to this one is relocated at the midpoint, and the short vector itself is deleted. The second case is treated by a procedure that measures the angle of intersection at vertices where exactly two vectors meet. If the angle is less than a predefined value (35 degrees by default), the vertex is removed. This correction phase is, however, incapable of solving bond–character touching problems and broken line detection problems.

Component Classification

Vectors produced by the GIFTS vectorizer are assembled into connected groups. These groups are classified as characters, bond structures, or other symbols such as circles representing aromatic rings. This is done using the size of each group as follows:

- If the ratio of the maximum dimension of the group to the maximum dimension of the diagram is less than a preset value, then that group is considered to constitute a symbol rather than a portion of the diagram line structure.
- Small groups containing only a few vectors are classified by position context: If located close to another letter, they are classified as characters. This rule accommodates the occurrence of lowercase l, as in the chemical symbol for chlorine, Cl, which could be mistaken for a bond on the basis of shape alone.
- If the group has at least N vectors ($N = 8$ by default) and is *circular*, a property measured in a special routine, then the group is classified as a circle. A similar group with fewer vectors is processed as a bond structure.
- If the group satisfies none of the above classifications, then it is classified as a bond structure.

OCR

First, the image of the character is extracted from the page bitmap. This process also attempts to detect and separate touching characters. Extracted patterns are subjected to size normalization and sent to a single-font OCR process. The OCR function uses feature-based software as referenced in Itoh and Takahashi (1990) to interpret character images.

Building a Connection Table

A connection table is built by sequencing through the groups and adding each character to the table of atoms. The bond structures are also processed, and a carbon atom is associated in the table at any point where two lines connect. The procedure then finds any vectors that have not yet been connected and associates them to the nearest atom or letter.

The identified character strings (e.g., CH_3) are parsed to map them into connection tables and detect their points of attachment to the rest of the structure. If a string cannot be associated with a node of the structure, the process looks for another string located above or underneath it. A string with no connection table is ignored and deleted from the connection table.

For each object classified as a circle, an aromatic-ring construction method is invoked during which bonds surrounding the circle are identified and every second bond is converted to a double bond. Finally, there is a check for valency violation.

CLiDE

The CLiDE program (Ibison et al. 1993) establishes the connection table of a digitized chemical structure in five steps:

- Component classification based on contour search
- Vectorization
- Dashed-line construction using Hough transforms
- OCR
- Creation of a connection table

Component Classification

First, the bitmap image is segmented into connected components. A connected component is a connected on-pixel region of the image. Connected components are represented by their outermost and innermost on-pixel sequences, that is, by their external and internal contours (see Figure 4.11). The contours are defined as the coordinates of a starting point and a sequence of four directions (N, S, E, W).

Connected components are separated into five basic groups: noise, characters, dashes, lines, and graphics. The separation is implemented as a stepwise algorithm, which separates one group from the others in each step. This separation is based on size, aspect ratio, and on-pixel density. The process uses several parameters (e.g., the average height of the connected components, the estimated maximum height of characters) that are set automatically during the processing phase and based on the statistical data extracted from the image (Venczel 1993).

This method usually works well for a wide range of character point sizes and graphic sizes. However, some confusion arises when a character touches a graphical component (bond) and hence they are classified together as graphics or when a dash-like character (e.g., "l") is classified as part of a dashed line. These separation errors are corrected by component reclassification at later phases of processing, for example, the dashed-line detection or the connection table building phases.

Vectorization

First, the contour fractions are extracted by cutting the contours of the image into straight and curved fractions (see Figure 4.11c). For each contour, a polygon is created in such a way that each point of the original contour is within a certain distance of a side of the polygon. If this threshold value is well chosen, straight parts of the contour result in long polygon sides, whereas curved parts are approximated by consecutive short sides. A method similar to that of Sklansky and Gonzalez (Sklansky and Gonzalez 1980; Venczel 1993) is used to create the approximation polygon. Long polygon sides are selected as straight contour fractions, and consecutive short sides are merged into curved fractions. Individual short sides are not used henceforth.

Vectors are found by searching for pairs of fractions that are adjacent. In an ideal case, two fractions are created for each line-like object, which are the two borders. During this step, the errors of the fraction creation phase are corrected. If the image is noisy, the fraction detection phase can create more than two fractions for one line by incorrectly cutting one border at an internal point (see Figure 4.12). In such a case, both parts lie side-by-side, and an attempt can be made to join them. Alternatively, if the two parts cannot be joined, it indicates that they belong to different lines sharing the fraction on opposite sides, and the shared fraction is cut into two (see Figure 4.13). Finally, each vector is described by the two borders. The coordinates of the line end points, the line width, and the line shape are determined from these two borders.

Dashed-Line Construction

Dashed-line detection is based on the Hough transform. This method detects dashed lines by searching for sets of connected components, classified as dashes, situated equally spaced along a straight line (Venczel 1993). At the end of this process, dash

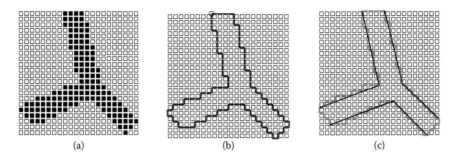

FIGURE 4.11 Contour determination in the CLiDE system. (a) Bitmap containing one connected component. (b) Contour of the connected component where the starting point of the contour is marked with a circle and the first few directions are displayed while following the contour in clockwise direction. (c) Straight fractions of the contour.

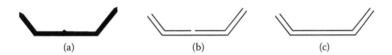

FIGURE 4.12 Joining of fractions during vectorization in the CLiDE system. (a) Connected component with three straight lines, with noise occurring in the center of the middle line. (b) Straight fractions detected; note that there are two fractions for the top border of the horizontal line. (c) Straight fractions after joining the two fractions belonging to the same lower border of the horizontal line.

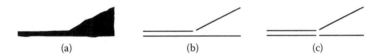

FIGURE 4.13 Cutting of fractions during vectorization in the CLiDE system. (a) Connected component of a straight line and a wedged line. (b) Straight fractions detected; note that there is only one fraction for the bottom border of the line and the wedge. (c) Straight fractions after cutting the fraction belonging to both lines.

components not belonging to any detected dashed line are reclassified as characters, lines, or noise according to their size and location compared to their surrounding connected components.

OCR

To date, two different OCR methods were implemented in CLiDE. The first one used a back-propagation neural network for classification of the characters. The character features used as input to the neural network are determined by template matching (Venczel 1993). The second OCR implementation in CLiDE is based on topological and geometrical feature analysis, and it uses a filtering technique for the classification of characters (Simon 1996).

Recognized characters are then grouped to form words based on their coordinates. This method combines characters lying next to each other but also considers vertical relationships to handle vertically oriented atom labels.

Creation of Connection Tables

Connection tables are built from the recognized solid and dashed lines and atom labels (Kam 1994). A bond line is represented by its two end points and a *free-flag* associated to each end. The free-flag is set to false if a bond line is touching another bond line with its end to which the free-flag is associated; otherwise, it is true.

First, it is determined that bond lines are connected to the atoms. A bond line end is joined to an atom if (a) the gap between them is smaller than a certain threshold, (b) the free-flag is true for the bond end in consideration, and (c) the bond points toward the atom label (see Figure 4.14). Next, bond ends not connected to any atom are joined together to form implicit carbon atoms. The main criterion for joining two bond ends is that they lie close to one another. There are potential difficulties in the case of crossing bonds, but the algorithm deals successfully with those problems. Finally, this connection information is converted to a connection table.

Atom labels are identified according to a superatom database that contains all the elements of the periodic table, the most frequently occurring functional groups and labels commonly used to represent R-groups in generic structures (e.g., R, R_1, R_2, R', X, Y). The database provides information, such as the name, the nature, the atomic and id code, the connection table (for groups), and so on, for each item contained therein. Longer atom labels, that is, linear representations of structural formulas not found in the database (e.g., $CF_3CF_2CH_2$) are parsed (Simon 1996).

CHEMOCR

The chemoCR system (Algorri et al. 2007a, b; Zimmermann et al. 2005) extracts molecules from an image via the following steps:

- Preprocessing
- OCR and component classification into characters and noncharacters
- Vectorization
- Reconstruction of molecules using a library of chemical graph-based rules and a set of chemical knowledge rules

FIGURE 4.14 Example on joining between atoms and bonds in the CLiDE system. Although the atom label *OAc* is closer to the two solid bonds (*b* and *c*) than to the dashed bond (*a*), *OAc* is correctly joined to bond *a*.

Preprocessing

First, the chemical image, which can be a color or grayscale image, is binarized. For this, the system first eliminates the effects of any anti-aliasing filter that might have been applied to the image and then uses an adaptive histogram binarization algorithm. The user can choose to manually adjust the adaptation parameter of the binarization every time a new image corpus is being recognized in order to fine tune the degree of thinning allowed by the binarization. The image is then segmented into connected components using a nonrecursive connected component algorithm based on the technique of raster scanning the image and identifying connected RLE segments (Algorri et al. 2007c). An RLE segment is a set of adjacent *on*-pixels lying in the same row. A connected component is a set of neighboring RLE segments, none of which is touched by RLE segments of any other connected component.

OCR and Component Classification into Characters and Noncharacters

This is done by a chemically oriented OCR (Akle et al. 2007) that identifies isolated characters and symbols and gives a confidence value to the recognition. The OCR analyzes the images of the connected components and extracts their features by calculating their moments using wavelet functions. The final identification of characters and symbols is done with a support vector machine algorithm. The OCR is able to classify the connected components that constitute characters and discard those that constitute parts of the molecular structure. The support vector machine algorithm can dynamically increase its training corpus every time it fails to correctly identify a connected component, therefore avoiding repetition of classification mistakes. The OCR can be trained to identify letters, numbers, and symbols of different sizes and fonts and is tolerant to some degree of rotation.

Vectorization

The connected components classified as noncharacters by the OCR are vectorized to produce a graph of vectors. This is done using a custom vectorizer (Algorri et al. 2007c) that takes as entry the connected components formed of connected RLE segments and assigns a local direction to every RLE segment based on the position of the RLE segment with respect to its neighboring RLE segments. The vectorizer then groups the directed RLE segments into patterns of local directions. The patterns of local directions have a very good correspondence with the vectors (global directions) in the image. The vectorizer results in a graph of vectors (edges and vertices) that includes ordered neighborhood information; that is, it contains information about which vectors share a common vertex and how they are positioned with respect to the vertex (orientation in degrees) and therefore which vectors are connected to each other.

Reconstruction of Molecules

The reconstruction of molecules is performed in two phases. The results of the previous two steps (the characters and the graph of vectors) are first chemically annotated and interrelated to create a chemical graph that is turned into a chemical molecule in the second phase.

The chemical graph construction is governed by a library of rules that describe the chemical characteristics of the elements in a molecule in terms of their geometrical properties in an image. The library is constructed using a few basic geometric measurements over the graph of vectors: length of the vectors, ordered neighborhood information, orientation, position, and geometric measurements over the characters: size, aspect ratio and position of the bounding box, and neighborhood with respect to the other characters and the vectors. The chemical graph is constructed in three steps: (a) bond annotation, (b) identification of atom labels, and (c) construction of the chemical graph.

- **Bond annotation:** Every vector is annotated as one of six allowable bond types: single, double, or triple bond; wedged or dashed wedged bond; and cross bond. The *wedged bonds* represented in the image as thick structures, a triangle or a rectangle, are the first ones to be annotated. To identify these bonds, every vector in the graph is registered (superimposed) with the pixels from the image that generated it, and a statistic is created to see how every vector registers with its corresponding pixels. Using this statistic, vectors generated by lines are identified as opposed to vectors generated by thicker geometrical forms. Next, vectors representing *dashed bonds* are annotated. To identify dashed bonds, all the vectors are selected that are not connected to any other vector and have no neighbors. These vectors are clustered using a quadtree clustering technique over the geometrical center of vectors. The vectors inside the resulting clusters are tested for parallelism and size coherence, and, if accepted, the set of vectors is fused into one vector that is labeled as a dashed bond. *Double and triple bonds* are identified from the vectors that are not connected to any other vector and that were not identified as dashed bonds. A region of interest around every bond (the bounding box of the vector dilated by a factor of two is defined, and any other vector intersecting the region of interest is tested for parallelism and size coherence. Vectors annotated as double (triple) bonds result from the fusion of two (three) parallel vectors. More than three parallel vectors generate a dashed bond. The remaining bonds are annotated as *single bonds*, a subset of which form cross bonds. *Cross bonds* are annotated by identifying neighboring vectors (vectors that touch or intersect each other) that intersect each other along the center part of the vectors rather than at the endpoints.
- **Atom labels:** In this stage, results of the OCR are analyzed to form atom labels. To form atom labels containing more than one character, characters are clustered by dilating the bounding box of every identified character by a factor of two. Characters whose dilated bounding boxes touch each other are clustered together. The center points of the involved bounding boxes are tested to determine if the characters should be horizontally or vertically grouped. To solve any occurring ambiguities, for example, when two atom labels are too close together and all the characters are clustered into one group, a dictionary of valid atom and functional group names are used.
- **Chemical graph:** Here, atom labels are associated to the vertices of the vector graph, thus creating a chemical graph. The vertices in the vector graph correspond to the end points of the vectors and can be shared by more than one vector. To bind the atom labels to the graph vertices, their center of mass is associated to its closest vertex in the graph. Vertices not bound to any atom label are associated to a default carbon atom.

In the second stage of molecule reconstruction, the chemical graph is converted to a chemical molecular structure. The vertices in the graph become the atoms in the chemical molecular structure, and the graph edges become the bonds, and their chemical valences, charges, and properties are validated by a set of chemical knowledge rules (Karl 2007).

RECOGNIZED PROBLEMS IN STRUCTURE RECOGNITION

With any of the systems discussed, it is very likely that an incorrect connection table will be built if there are no specific rules to detect that a structure diagram contains a feature that is unusual or conveys an ambiguous situation. Some of these difficult features that have been identified are discussed below.

Crossing bonds are one of the kinds of difficult features of chemical drawings. They are often used to preserve some sense of the three-dimensional shape of the molecule in the drawing, particularly in bridged structures. A further reason for the presence of a crossing bond in a structure arises where the bond being crossed is a ring bond and the crossing bond indicates that the position of attachment of this bond to the ring is not specified. In the bond crossing situation, one bond usually "cuts" another, although there is often no apparent gap in the bond that is crossed. Figure 4.15 illustrates the different types of crossing bonds that can be found in chemical structures. To successfully interpret a structure, one has to detect the bond crossing situations, identify the bond crossing types, and construct a connection table according to these types. The crossing bond interpretation method developed in CLiDE (Kam 1994; Kam et al. 1992; Ibison et al. 1993) uses a set of rules, including the proximity, length, collinearity, and ring membership of potential crossing bonds, to correctly detect and interpret all the crossing bond types. Test reports show that the method works on a wide range of structure diagrams containing crossing bonds.

It is also difficult to handle connected components that cause ambiguity in interpretation. This mainly evolves from simple single lines. For example, a vertical

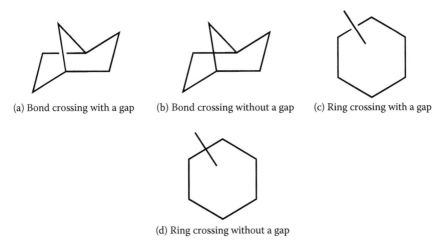

(a) Bond crossing with a gap   (b) Bond crossing without a gap   (c) Ring crossing with a gap

(d) Ring crossing without a gap

FIGURE 4.15   Different types of crossing bonds.

line can occur in several different kinds of chemical entities such as single and multiple bonds, dashed bonds, and character strings representing atom labels (e.g., Cl, I) and other information related to the structure. Similarly, a horizontal line can be part of a single or multiple bond or a dashed bond, or it can represent a negative charge for an atom. Such ambiguous situations can be resolved by analyzing the environment of the connected components and applying a set of rules with conditions on chemical and spatial context. For instance, if a letter C is on the left side of a vertical line that is not part of a dashed bond, the vertical line represents the letter l of a chlorine atom.

Some bond formations can be easily misinterpreted, and postprocessing of the interpreted connection table is needed to get correct results. Merely relying on the end points of the vectors calculated for each bond line, a single bond and a triple bond joined together (see Figure 4.16a) can be recognized as a long single bond and a double bond half way over the single bond. Figure 4.16b shows another bond formation requiring special treatment. Here, a broken line directly attached to an implicit carbon atom of a ring can be misinterpreted as a broken-line dashed bond and a short single bond with a carbon atom between the two bonds.

The performance of optical structure recognition is highly affected by the quality of the input image. Noise is one factor that can cause considerable deterioration of the image quality. One kind of noise, isolated small black spots often appearing during the scanning of documents, can be easily separated from the connected components of normal image objects and ignored. Bigger isolated noise is harder to identify and exclude from the structure recognition process. Black spots touching one or more connected components have a big deleterious impact on the structure recognition because they change the shape and the number of the connected components and they are very difficult to detect and isolate to retrieve the correct shape of the connected components. In some images, tiny white spots occur totally or partially inside connected components. White spots lying entirely in the black area of connected components can be easily detected in images containing relatively big connected components. White spots fusing with the background are generally difficult to detect and handle. White spots, if they are untreated, have a deleterious effect on OCR and vectorization.

(a) Triple bond joined to a single bond   (b) Broken-line bonds attached to solid bonds

FIGURE 4.16 Structure diagrams containing bond formations that can be easily misinterpreted.

RECOGNITION OF COMPLEX OBJECTS

RECOGNITION OF GENERIC STRUCTURES

Interpretation of generic, or Markush, structures is a very useful but nontrivial task, and from the published work, it appears that so far, CLiDE is the only system among the various structure recognition systems introduced in "Projects," above, that addresses the problem of generic structure interpretation.

The extraction of generic information from structures can be divided into two parts. First, the generic text block needs to be identified and extracted from other text blocks in the text region and its meaning interpreted. This is done by a generic text interpreter (Ibison 1992) whose task is to extract the generic information from the text. The stages involved in this task include determining the R-groups, the number and type of the substituents, and whether any label is present for each substituent. The generic text interpretation in CLiDE (Simon 1996) is performed in three phases: (1) lexical analysis or tokenization that isolates the individual words (or tokens) of the sentences, (2) syntax analysis that identifies the parts of the sentence requiring contextual checking, and (3) semantic analysis that determines the meaning of the words. One of the limitations of the current CLiDE generic text interpreter is that it relies on the presence of special symbols such as an equals sign separating R-groups and substituents, and delimiters. A universal generic text interpreter is required, because the generic text block can appear in different formats. For instance, the alternatives of R-groups can be expressed in various ways, mostly by chemical symbols and formulas, as well as chemical drawings (see Figure 4.17). When chemical drawings are used, they have to be identified and interpreted by a graphical recognition module. Furthermore, R-groups and their substitution values can be listed in a table (without any equals signs) rather than in a linear text format.

After the information about each generic text block has been interpreted, the choice of the most suitable generic text block for a given structure has to be made. In CLiDE, this is done in two stages. First, a search is performed to find the generic text blocks that best match the structure in terms of the number of R-groups present in both the structure and the generic text block. In the second phase, the generic text block lying closest to the structure is selected from those found in the first phase. A match between a structure and a generic text block goes further than checking whether an R-group in a structure matches an R-group in the text. For example, in the structure, an atom label might be OR, whereas the R-group in the text might be

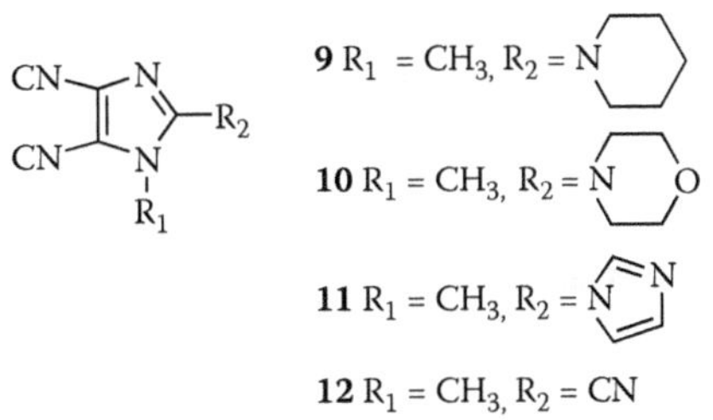

FIGURE 4.17   Generic substitutions represented by chemical graphics.

R, which is just part of OR. A simple comparison between R and OR is not enough; instead, it is the generic element in a generic type that is important. The generic element in both OR and R is R. Therefore, "match" means that the generic elements in the generic atom labels of the structure and the generic elements in the generic text block are the same. Figure 4.18 depicts a generic structure that can be interpreted by the new version of CLiDE, CLiDE Pro, by successfully identifying the R-groups X, Y_1, Z, and R in the text, recognizing the R-group substitution values N, CF, H, Boc, and Et, and detecting the generic elements in the text and in the generic atom labels of the structure (X, Y_1, ZHN, CO_2R) to find a match between the text and the structure. However, this process heavily relies on the success of the individual steps performed; if one step fails, automatic error detection and recovery in the subsequent steps are difficult without human intervention.

EXTRACTION OF REACTION INFORMATION

Although reaction schemes containing graphical structures are frequently used in chemistry-related documents, CLiDE is the only system that currently attempts extraction of reaction information.

In CLiDE (Kam 1994), the interpretation of reactions is performed in two phases. In the first phase, the reaction scheme primitives, arrows, the reagent text, and the joiner "+" are identified. The lines in structures and the lines in arrows are initially gathered. During arrow extraction, this original set is divided into two groups of items: a structure lines set and an arrow lines set. The structure lines are interpreted as bonds during the compilation of chemical graphs. The arrow lines are constituted into arrows, and the arrow types are determined (see Table 4.1, which lists common arrow types). This process considers that all the arrows in the scanned image are reaction arrows and ignores the case where an arrow is part of a structure (see Figure 4.19). Once the arrows are found, the text groups in the image are then checked to see whether they constitute reagent text relating to an arrow. If they do, they are appended to the arrow and give more complete information about the reaction. The method of finding reagent text for an arrow is divided into two parts. The first part finds text strings that are directly next to the side of the arrow. The second part uses the positions of these text strings to find other text strings that are near them. The reason for the second part is that a reagent text may contain more than one

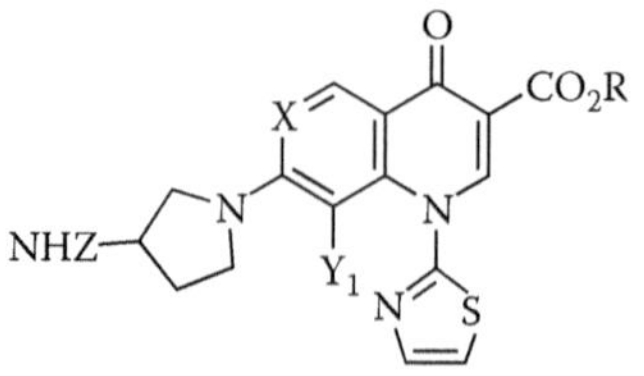

42a: X = N, Y_1 = H, Z = Boc, R = Et
42b: X – Cl, Y_1 – H, Z – H, R – Et
42c: X = CF, Y_1 = F, Z = Boc, R =H

FIGURE 4.18   A generic structure interpretable by CLiDE Pro.

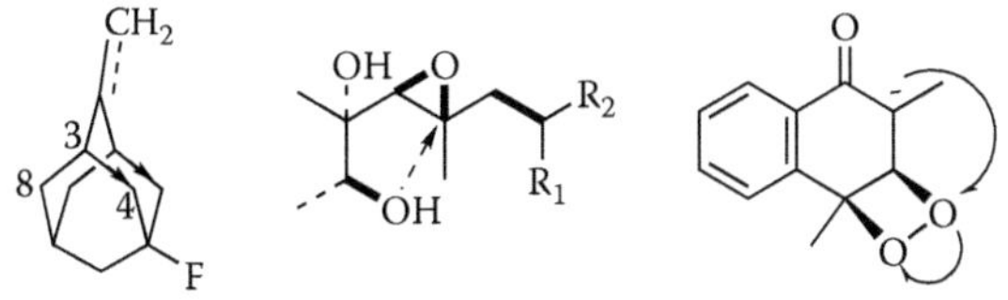

FIGURE 4.19 Arrows as part of structure diagrams.

line, and some lines are quite far from the arrow but should still be near the other reagent text of that arrow. After the detection of arrows and reagent text, the connection tables are constructed. Any positive charge "+" for an explicit atom or functional group or for an implicit carbon atom should already have been found; therefore, any + sign left is assumed to be a potential joiner between structures.

In the second phase of reaction interpretation, reactants and products are extracted and their roles assigned. Based on the relative positions of the joiner "+" and the structures, it is first determined whether the structures should be considered as one structure group. Next, the set of reactants and the set of products are allocated for each reaction in a scheme by finding structures that lie behind an arrow tail or in front of an arrow head and are passed by the line going along the arrow. This process considers both single structures and groups of structures and can interpret different reaction formats including situations where reactants and products are placed horizontally, vertically, or diagonally. Line breaks are also handled where an arrow is located at the margin of a page and because of the spatial limitation of a page, the products and reactants are not on the same row as the arrow. A reaction containing a complex arrow with a number of arrow heads or arrow tails is split into separate simple reactions by considering each arrow head and arrow tail like that of a simple arrow, and for each will be found its own set of products and its own set of reactants. Graphic reagents are identified as additional reactants of a simple reaction by finding structures that are near the arrow and situated either above, below, left, or right of the arrow. Figure 4.8 shows a three-step reaction scheme containing a joiner "+" between the reactants of the first reaction, two line break situations, and a graphic reagent.

OUTPUT DATA

The primary aim of optical structure recognition is to convert printed chemical structure diagrams into connection tables and other computer-readable formats suitable for chemical structure database updating and searching. Therefore, the output produced by the systems introduced in this chapter is available in standard chemical formats such as Molfile and SMILES. However, these proprietary formats contain only the raw molecular structure information that their design allows. The SMILES file contains the connection table of a structure, so page numbers, titles, and associated text and images are lost in converting to this format. For this reason, all information that has been extracted by a recognition software tool is saved in its own internal file format, which can be much "richer" than the standard formats.

CONCLUSION AND OUTLOOK

The prospects of automatic conversion of chemical structure images into computer-readable format are quite attractive and, consequently, this field has attracted computer scientists working on this problem since the early 1990s. However, the various scientific challenges behind optical structure recognition are substantial, and the vision of large-scale automated interpretation of chemical information from large digital archives will require sustained research and development over a number of years.

The most significant bottleneck of optical structure recognition is the high dependency on the image processing algorithms. Errors in the image processing phase often result in failure of the recognition process to correctly identify a structure. Errors in the binarization process (see "Input Data," above) impact all succeeding image processing steps. If a wrong binarization threshold is selected, too few border pixels of image objects are removed and two normally distinct objects will remain in one connected component. For example, the character O might be included in a connected component representing a bond set, and the OCR cannot recognize the character, and the vectorized bond set holds strange vectors. Errors in the reconstructed molecule can also occur when characters are not recognized properly. Uncertainty about the correct classification of character symbols is a common problem in the field of OCR. Uncertainty often happens where characters cannot easily be distinguished based on their shape; for example, the characters l, 1, (,'), |, and I can cause confusion in the classification process. In ordinary text, an OCR tool would effectively address this problem by applying diverse postprocessing steps. A well-established technique is to verify the concatenated string by searching for the recognized word in a dictionary. It is accepted if the reference book contains it or a similar word is found. However, the strings in a molecular structure are usually very short (usually just one or two characters), so there is no contextual information like that seen in the recognition of normal text documents. Errors produced by the vectorization algorithm have a big deleterious effect on the construction of a connection table. A wrong identified character does not change the overall appearance of the molecule, whereas a wrong bond can influence the entire topology. Both OCR and vectorization can be improved by incorporating chemical knowledge into the recognition of characters and bonds. The number of errors can be reduced by extensive error checking throughout the recognition process.

Though far from perfect, at least some of the various recognition methods described in this chapter can yield impressive recognition results. Figure 4.20 shows some structure diagrams that are correctly interpreted by optical structure recognition within a few seconds. Manual reproduction of these structures would require considerably more time and be error prone.

The optical structure recognition tool that has 100% accuracy in all situations has yet to be developed, and indeed is unlikely to be developed in the foreseeable future. This parallels the situation for text OCR, where despite decades of research, accuracy of recognition still falls short of 100%. However, this is perfectly acceptable as long as the extent of manual editing required is sufficiently small that the overall process provides significant time savings over manual transcription. This is also true of optical structure recognition. Some of the existing tools are capable of very high

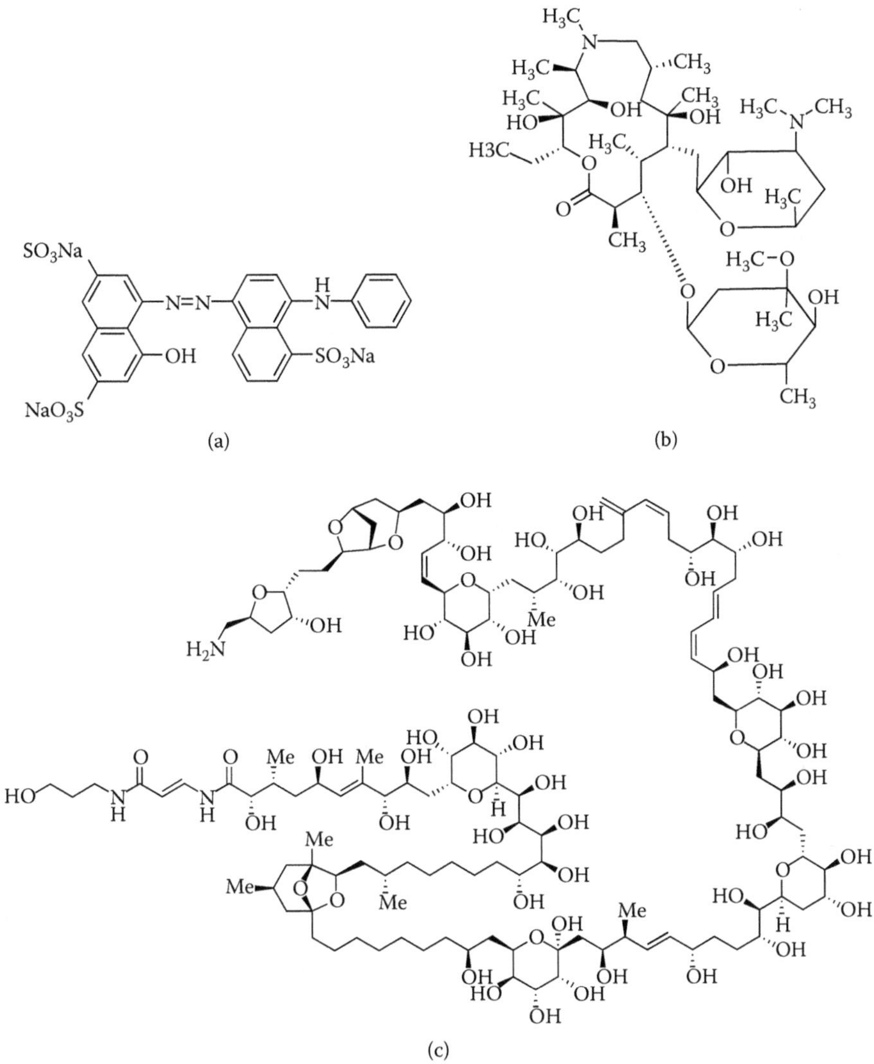

FIGURE 4.20 Structure diagrams correctly interpreted by (a) Kekulé, (b) chemoCR, and (c) CLiDE Pro.

accuracy when used with good-quality images and could be integrated into work-flow processes, with minimal manual editing required at the end of the process, thus providing enormous time savings over manual redrawing. Poorer-quality images, particularly those from older volumes of journals where the structures were drawn manually, are likely to give rise to more errors, and improving accuracy for these cases is an important area of research in this field.

It should be recognized that structure recognition on its own is only a part of the problem of extracting chemical information from documents. The data associated with structures needs to be captured, as does the context of the structure, for example, the reactant or product of a reaction or a Markush structure with associated values of R groups. The CLiDE system has already made considerable headway in this area, but there is still much to be done, and this field is likely to be the focus of intense research effort in the next few years.

It should be recognized that automated name recognition provides a realistic alternative to optical structure recognition in many cases. Indeed, the two techniques are complementary; there are many cases where either no name is given (particularly common with complicated structures) or the name is insufficiently precise to convey the full structural information (including stereochemistry). The ideal system would incorporate both techniques and use either whichever is more appropriate in a given situation, or even both in parallel to provide an accuracy check.

Diverse text mining approaches have been successfully applied to the scientific literature in biology and medicine. The situation is less well developed for the chemical literature because of the focus on compound structures rather than text. The advent of optical structure recognition offers the promise that mining the chemical literature will be even more successful than it is in these other fields of scientific endeavor.

REFERENCES

Aho, A. V., J. E. Hopcroft, and J. D. Ullman. 1983. *Data Structures and Algorithms*. Reading, MA: Addison-Wesley.

Akle, S., M. E. Algorri, C. Friedrich, and M. Zimmermann. 2007. *Chemical OCR*. Fraunhofer SCAI Internal report.

Algorri, M. E., M. Zimmermann, and M. Hofmann-Apitius. 2007a. Automatic recognition of chemical images. *Eighth Mexican International Conference on Current Trends in Computer Science*, (ENC 2007):41–46.

Algorri, M. E., M. Zimmermann, and M. Hofmann-Apitius. 2007b. Reconstruction of chemical molecules from images. Engineering in Medicine and Biology Society, 2007. EMBS 2007. *29th Annual International Conference of the IEEE*, pp. 4609–4612.

Algorri, M. E., M. Zimmermann, M. Hofmann-Apitius, and C. Friedrich. 2007c. *Turning a Binary Image into a Vector Graph by Analyzing the Texture of Discrete Directions*. Fraunhofer SCAI Internal report.

Baird, H. S. 1994. Background structure in document images. *Int. J. Pattern Recognit. Artif. Intell.* 8:1013–1030.

Barnard, J. M., M. F. Lynch, and S. M. Welford. 1981. Computer storage and retrieval of generic structures in chemical patents. 2. GENSAL, a formal language for the description of generic chemical structures. *J. Chem. Inf. Comput. Sci.* 21(3):151–161.

Barnard, J. M., M. F. Lynch, and S. M. Welford. 1982. Computer storage and retrieval of generic structures in chemical patents. 4. An extended connection table representation for generic structures. *J. Chem. Inf. Comput. Sci.* 22(3):160–164.

Borman, S. 1992. New computer program reads, interprets chemical structures. *Chem. Eng. News* 70(4):17–19.

Casey, R., S. Boyer, P. Healey, A. Miller, B. Oudot, and K. Zilles. 1993. Optical Recognition of Chemical Graphics. *Proc. 2nd Int. Conf. on Doc. Anal. and Recogn.* (ICDAR'93), pp. 627–631.

Chowdhury, G. C., and M. F. Lynch. 1992a. Automatic interpretation of the texts of chemical patent abstracts. 1. Lexical analysis and categorization. *J. Chem. Inf. Comput. Sci.* 32(5):463–467.

Chowdhury, G. C., and M. F. Lynch. 1992b. Automatic interpretation of the texts of chemical patent abstracts. 2. Processing and results. *J. Chem. Inf. Comput. Sci.* 32(5):468–473.

Contreras, M. L., C. Allendes, L. T. Alvarez, and R. Rozas. 1990. Computational perception and recognition of digitized molecular structures. *J. Chem. Inf. Comput. Sci.* 30(3):302–307.

Duda, R. O., and P. E. Hart. 1972. Use of the Hough transform to detect lines and curves in pictures. *Graphics Image Process.* 1 (2):409–418.

Fan, K. C., C. H. Liu, and Y. K. Wang. 1994. Segmentation and classification of mixed text/graphics/image documents. *Pattern Recogn. Lett.* 15(12):1201–1209.

Fisanick, W. 1990. The Chemical Abstracts Service generic chemical (Markush) structure storage and retrieval capability. 1. Basic concepts. *J. Chem. Inf. Comput. Sci.* 30(2):145–154.

Fletcher, L. A., and R. Kasturi. 1988. A robust algorithm for text string separation from mixed text/graphics images. *IEEE Trans. Pattern Anal. Mach. Intell.* 10(6):294–308.

Gkoutos, G. V., H. Rzepa, R. M. Clark, O. Adjei, and H. Johal. 2003. Chemical machine vision: automated extraction of chemical metadata from raster images. *J. Chem. Inf. Comput. Sci.* 43(5):1342–1355.

Govindan, V. K., and A. P. Shivaprasad. 1990. Character recognition: a review. *Pattern Recognit.* 23(7):671–683.

Ibison, P. 1992. *Internal Report in Image Group.* Maxwell Institute.

Ibison, P., M. Jacquot, F. Kam, et al. 1993. Chemical literature data extraction: the CLiDE Project. *J. Chem. Inf. Comput. Sci.* 33(3):338–344.

Ibison, P., F. Kam, R. W. Simpson, C. Tonnelier, T. Venczel, and A. P. Johnson. 1992. Chemical structure recognition and generic text interpretation in the CLiDE Project. *Online Inf.* 92:131–142.

Illingworth, J., and J. Kitter. 1988. The adaptive Hough transform. *Comput. Vision Graphics Image Process.* 44(2):87–116.

Itoh, N., and H. Takahashi. 1990. A handwritten numeral verification method using distribution maps of structural features. *Proc. SPIE Image Communications and Workstations*, p. 1258.

Ittner, D. J., and H. S. Baird. 1993. Language-free layout analysis. *Proc. 2nd Int. Conf. on Anal. and Recogn.* (ICDAR'93), pp. 331–340.

Jain, A., and S. Bhattacharjee. 1992. Text segmentation using gabor filters for automatic document processing. *Machine Vision Applications* 5:169–184.

Kam, F. 1994. Automated Extraction of Chemical Information from the Chemical Literature. PhD Dissertation, School of Chemistry, University of Leeds, U.K.

Kam, F., R. W. Simpson, C. Tonnelier, T. Venczel, and A. P. Johnson. 1992. Chemical literature data extraction. Bond crossing in single and multiple structures. *Int. Chem. Inf. Conf.*, pp. 113–126.

Karl, P. 2007. Chemical structure recognition via an expert system guided graph exploration. Master's thesis, Ludwig-Maximilians-Universitaet, Munich, Germany.

Krishnamoorthy, M., G. Nagy, S. Seth, and M. Viswanathan. 1993. Syntactic segmentation and labeling of digitized pages from technical journals. *IEEE Trans. Pattern Anal. Mach. Intell.* 15(7):737–747.

Loening, K. L. 1988. Poster session: conventions, practices, and pitfalls in drawing chemical structures. *Chem. Structures Int. Lang. Chem.*, pp. 415–423.

Lynch, M. F., J. M. Barnard, and S. M. Welford. 1981. Computer storage and retrieval of generic structures in chemical patents. 1. Introduction and general strategy. *J. Chem. Inf. Comput. Sci.* 21(3):148–150.

McDaniel, J. R., and J. R. Balmuth. 1992. Kekule: OCR-optical chemical (structure) recognition. *J. Chem. Inf. Comput. Sci.* 32(4):373–378.

Naccache, N. J., and R. Shinghal. 1984. An investigation into the skeletonization approach of Hildtich. *IEEE Trans. Syst., Man Cyber* 3(2):409–418.

Nagy, G., S. Seth, and M. Viswanathan. 1992. A prototype document image analysis system for technical journals. *Computer* 25:10–22.

O'Gorman, L., and R. Kasturi. 1993. The document spectrum for page layout analysis. *IEEE Trans. Pattern Anal. Mach. Intell.* 15(11):1162–1173.

Pavlidis, T. 1968. Computer recognition of figures through decomposition. *J. Inf. Control* 12:526–537.

Saitoh, T., T. Yamaai, and M. Tachikawa. 1994. Document image segmentation and layout analysis. *IEICE Trans. Inf. Systems* 77(7):778–784.

Simon, A. 1996. Image Analysis and Content Retrieval of Printed Chemistry Documents. PhD Dissertation, School of Chemistry, University of Leeds, U.K.

Simon, A., and A. P. Johnson. 1997. Recent advances in the CLiDE Project: logical layout analysis of chemical documents. *J. Chem. Inf. Comput. Sci.* 37(1):109–116.

Simon, A., J. C. Pret, and A. P. Johnson. 1995. (Chem)DeTeX Automatic generation of a markup language description of (chemical) documents from bitmap images. *Proc. 3rd Int. Conf. on Doc. Anal. and Recogn.* (ICDAR'95), pp. 458–461.

Sklansky, J., and V. Gonzalez. 1980. Fast polygonal approximation of digitized curves. *Pattern Recogn.* 12:327–331.

Smith, R. W. 1987. Compuer processing of line images: a survey. *Pattern Recogn.* 1(2):7–15.

Tetsuo, A., D. Z. Chen, N. Katoh, and T. Tokuyama. 1996. Polynomial-time solutions to image segmentation. *Proc. Seventh Ann. ACM-SIAM Symp. Discrete Algorithms*, pp. 104–113.

Tsujimoto, S., and H. Asada. 1992. Major components of a complete text reading system. *Proc. IEEE* 80(7):1133–1149.

Venczel, T. 1993. Recognition of Primitives. PhD Dissertation, School of Chemistry, University of Leeds, U.K.

Zimmermann, M., T. Bui Thi Le, and M. Hofmann-Apitius. 2005. Combating illiteracy in chemistry: towards computer-based chemical structure reconstruction. *ERCIM News* (60):40–41.

5 Chemical Entity Formatting

Bedřich Košata

CONTENTS

INTRODUCTION

In this chapter, we will describe how computers store and process chemical structures. Our main goal will be to provide the reader with enough information to be able to choose from the variety of available chemical formats and implement the right methods for online publication of his or her data.

We will give an overview of the most commonly used chemical formats and their features, strengths, and weaknesses. We will also discuss several typical scenarios of online publication of chemical structural data and the closely related topic of storage of such data.

With a few exceptions, we will focus our attention only on the semantic part of the description of chemical structures and disregard the presentational aspects of a chemical drawing.

COMPUTER REPRESENTATION OF MOLECULES

The whole concept of modern chemistry is based on the usage of drawings or other visual models of molecular structures. These simplified models of real molecules provide us with a framework that helps us understand the behavior of chemical compounds. With the development of computers, scientists had to face the problem of transferring molecular models into computer language.

Even though it is possible to use the same kind of models in computers as humans use, the internal workings of a computer are very different from the human cognitive system. Whereas humans process a chemical drawing in a visual manner and are able to quickly identify important parts of a drawing, computers notoriously have big problems doing the same. In this sense, we can compare the task of visualizing a molecular structure by using its drawing to viewing a photograph. For humans it is a trivial task to recognize that a picture contains two children, one in red and one in blue jeans. However, for computers, such a task is immensely complicated (the topic of computer processing of chemical structures in the form of graphics is discussed in Chapter 4). Therefore, we cannot use pictures of chemical structures for efficient computer processing, and we need to provide the computer with a much more explicit description—a description that represents the model itself, not its visualization for human convenience.

Thus, in conventional computer models, molecules are represented directly as networks of atoms connected by bonds. (Figure 5.1 demonstrates this approach.)

This is a very natural way to describe molecular structures because their most important feature for most applications, the connectivity of atoms, forms the basis of the internal representation. The big advantage of this model is that it transforms many problems of chemistry into problems of network topology.

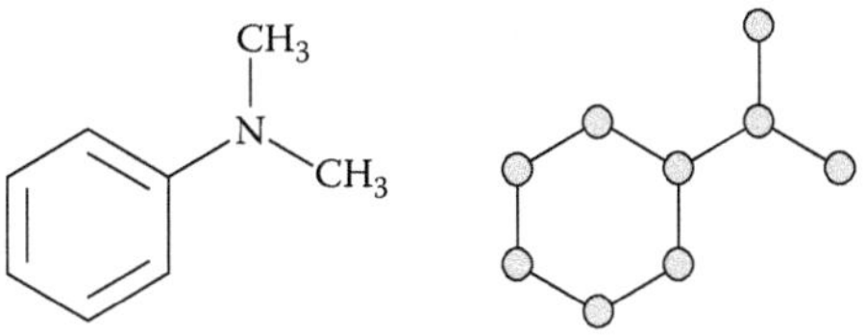

FIGURE 5.1   Model of a chemical structure as a network of interconnected nodes — a graph.

A general name for such a network as demonstrated in Figure 5.1 is a *graph*. Graphs are very well-studied objects that can be used to describe many nonchemical problems, for example, an electric grid, social connections of people, citations between scientific articles, and links between web pages. Because of this broad range of uses, *graph theory,* a branch of mathematics that studies graphs, is very well developed and has a solid mathematical background (for general introduction into graph theory see, for instance, Trudeau 1993 or Chartrand 1984; for an overview of graph theory in chemistry see, for example, Trinajstic 1992). Thus, many general graph algorithms can be used in treating chemical compounds, without the need to reinvent them for use in chemistry.

Because of the graphic nature of most molecular representations, the formats used to store such information are mostly oriented toward description of this graph. While this fact is more pronounced in some formats and less obvious in others, a graph is always hidden inside the format.

In the most general sense, a graph represents a network of nodes (called *vertices*, singular *vertex*) connected by links (called *edges*). As such, it can only describe the connectivity of atoms, not their other properties, such as the number of protons, electronic configuration, and so on, or the properties of the molecule as a whole—its chirality and so on. These features have to be added on top of the graphic representation of the molecule.

The graphic representation also does not handle and is not influenced by the positions of atoms. This means that many basic algorithms for analysis and manipulation of molecular graphs work independently of the positions of atoms and even without the positions being specified. This fact makes it possible to ignore molecular geometry during processing and even in storage, thus enabling creation of very compact chemical formats, such as the Simplified Molecular Line Entry System (SMILES) or International Chemical Identifier (InChI).

Of course, many problems also exist for which the actual geometry of a molecule has to be taken into account. In such cases this information has to be added on top of the basic model, because graph theory itself is not concerned with it.

BASIC OVERVIEW OF CHEMICAL FORMATS

Before we deal with the most common formats in more detail, we will try to classify them according to several practical rules to discuss some general topics involved.

From the point of view of computer storage and user interaction, we can distinguish two basic types: linear (or inline) and file-based formats. The former are designed to be used directly as part of text (an article, web page, etc.), and the latter use separate files for storage. This difference has a strong influence on the properties of corresponding formats. Linear formats tend to be compact so as not to clutter the actual text. Thus, they seldom store atomic coordinates and often omit other information about the structure. In contrast, file-based formats are not limited by the same factor and usually allow storage of the complete information about a molecular structure and often even have support for additional data storage. Thus, the decision to use either a linear or file-based format depends on the problem we face. Linear formats are simpler to use inside text and are much more space efficient; they are, however, completely unusable for visualization of computed or measured geometries and other similar problems.

In the group of linear formats, we can distinguish two basic subtypes: canonical and noncanonical. Canonical formats always assign one molecule the same code, regardless of how it was drawn—upside down, from left to right or vice versa, in two different conformations, and so on. Noncanonical formats allow more transcriptions of one structure. Although this may seem to be a clear setback, it has the advantage of allowing humans to write such formats by hand, without the need to understand the very complex algorithms that are involved in producing canonical output. Creation of canonical representations is, however, always done with help of computer programs. The basic part of the canonization process lies in canonical numbering of atoms, followed by encoding of the structure based on the obtained numbers—usually from the lowest one. The process may also optionally include a normalization step that transforms several chemically equivalent forms of one compound to one normalized structure.

Even with today's high availability of computers and specialized software, it is still important to distinguish between formats that are readable (and writable) by humans and those that can only be properly used with the aid of a computer. With the appearance of InChIKey, we now even have a widely supported format that is readable neither by humans nor computers because it contains only a "fingerprint" of a molecular structure. Such a format is in this respect similar to registry numbers (such as CAS RN or Beilstein RN) that cannot be deciphered without a database lookup. We will discuss this feature in more detail in the section describing InChIKey.

From yet another point of view, we can distinguish between formats that are designed only for the specific purpose of capturing chemical entities and formats that are capable of storing all aspects of a chemical drawing, such as line widths and font settings. Because of the need to share chemical information between scientists, there are already several standard and commonly accepted formats of the former kind. However, storage of complete chemical drawings including presentational aspects does not have such an important scientific value, even though it plays an important part in the publication process (where it can, however, be replaced by export to common graphical formats). The formats also need to be much more complex to allow for storage of all the necessary data. For these reasons, formats with support for presentation are usually limited to native formats of individual drawing programs, and there is no really widely accepted standard (even though some formats, such as the ChemDraw CDX and CDXML formats (http://www.cambridgesoft.com/services/documentation/sdk/chemdraw/cdx/) are more common than others). Because our interest in this chapter is in pure chemistry and in the most commonly used formats, we will discuss only those from the first category.

COMMON CHEMICAL FORMATS

In the previous section, we showed that widespread chemical formats fall into the category of formats intended only for description of chemistry, not for presentational purposes. In this section, we will focus in more detail on several such formats. An overview of the selected formats is given in Table 5.1.

MOLFILE AND RELATED FORMATS

Basic Characteristics

Molfile is a file-based format that:

- Features a simple text based format
- Contains molecular geometry
- May contain other related data
- Has very good software support

A format example (Molfile of ethanol) is as follows:

```
  3  2  0  0  0  0  0  0  0  0999 V2000
  146.0000  116.7456    0.0000 C   0  0  0  0  0  0  0  0  0  0  0  0
  168.9574  130.0000    0.0000 C   0  0  0  0  0  0  0  0  0  0  0  0
  191.9147  116.7456    0.0000 O   0  0  0  0  0  0  0  0  0  0  0  0
  1  2  1  0  0  0  0
  2  3  1  0  0  0  0
M  END
```

TABLE 5.1

Overview of Described Common Chemical Formats

Format	Linear or File Based	Human Readable[a]	Computer Readable
Molfile	File based	No	Yes
CML	File based	No	Yes
SMILES	Linear	Yes	Yes
InChI	Linear	No	Yes
InChIKey	Linear	No	No (write only)

[a] By human readability, we mean if the format was designed to be *easily* read by humans. Most of the computer-targeted formats are readable by humans who have deep enough knowledge of the format and plenty of time on their hands.

Format Overview

Molfile is probably the most common chemical format found on the Internet. Even though Molfile is in fact only the simplest member of a family of related formats known as CTFile formats (Dalby et al. 1992) developed by MDL (Molecular Design Limited), the name Molfile is often used to refer to all the CTFile formats in general.

All these formats are based on a connection table (CT; from this comes the name CTFile formats) for description of a molecular structure and differ in their capabilities and purpose. Table 5.2 gives an overview of the most commonly found CTFile formats.

While support for the other family members differs considerably from program to program, support for its simplest member, the Molfile, is found in almost all chemical drawing software available and in many other chemistry-related programs. This fact makes Molfile often the format of choice when chemical data are transferred, especially on the scale of individual compounds. On a larger scale and when additional data need to be attached, SDFiles or other more complex CTFile formats are often used.

CML (CHEMICAL MARKUP LANGUAGE)

Basic Characteristics

CML is a file-based format that:

- Features an XML-based format
- Contains molecular geometry
- May contain other related data
- Has moderate software support

TABLE 5.2
The CTFile Format Family

Name	Molecular Structure	Reaction	More than One Structure or Reaction	Description
Molfile	Yes	No	No	Simple format for description of one molecule (which may be disconnected)
Rnxfile	Yes	Yes	No	Format containing simple Molfiles describing one reaction
SDfile	Yes	No	Yes	Multistructure format used for storage and transfer of data
RDfile	Yes	Yes	Yes	Similar to SDfile but may contain reactions
XDfile	Yes	Yes	Yes	XML-based format for storage and transport of reaction and structure data with associated information

---

A format example (CML of ethanol) is as follows:

```
<?xml version="1.0" ?>
<cml>
 <molecule id="m1">
 <atomArray>
 <atom elementType="C" id="a2" x2="146.0" y2="130.0"/>
 <atom elementType="C" id="a1" x2="169.0" y2="116.7"/>
 <atom elementType="O" id="a3" x2="192.0" y2="130.0"/>
 </atomArray>
 <bondArray>
 <bond atomRefs2="a2 a1" order="1"/>
 <bond atomRefs2="a1 a3" order="1"/>
 </bondArray>
 </molecule>
</cml>
```

Format Overview

CML (Murray-Rust and Rzepa 1999, 2003) is an XML-based format with basic characteristics very similar to those of CTFiles when it comes to description of chemical structures. There are two main published versions of CML: CML 1 and CML 2. As the former one is already considered obsolete, we will describe only the properties of CML 2.

CML is most widely used to describe individual molecules, but is also capable of describing chemical reactions (Holliday et al. 2006), spectra (Kuhn et al. 2007), computational chemistry results and crystallographic data. Although only one specification of CML contains all the mentioned types of data, specific subsets of CML are informally referred to by a specific name (http://cml.sourceforge.net/wiki/index. php/FAQ), such as CMLReact for reactions or CMLSpect for spectra.

The decision of CML's authors to base a new chemical format on XML was very providential, even though at the time of CML's creation XML did not have today's general acceptance. In the meantime XML has become widely supported, and many other formats were based on its foundations.

One of the most important benefits that XML brought to CML is the existence of standard validation tools, which, equipped with definition of the language, can automatically check the formal structure of a CML document and report possible inconsistencies. For XML, there is a wide variety of generic software tools and libraries that enable checking, reading, and transformation of documents. Even though these tools are not directly focused on CML, they represent a good foundation for manipulation of the format. More about the benefits of XML and CML itself can be found in Chapter 6.

Because CML is much more recent that CTFile formats, it does not have such a broad acceptance and software support. Another reason for this state is also probably the fact that CTFile formats suit their purpose very well and there is not much need to replace them with a new alternative.

However, XML is constantly penetrating farther into all fields of computer science, and it is probably only a matter of time until CML or a similar format replaces Molfile as the de facto standard format for molecular structure storage.

SMILES

Basic Characteristics

SMILES is a linear format that:

- Has human-writable and -readable format
- Optionally may have canonical form
- Has good software support

A format example (SMILES of ethanol) is as follows:

CCO

Format Overview

The name SMILES comes from Simplified Molecular Input Line Entry System (Weininger 1988). It is a linear format for description of molecules and reactions developed by Daylight Chemical Information Systems, Inc. SMILES was designed with the intention to be both human readable and writable, which makes it unique among the other chemical formats described in this chapter.

Like Molfile, SMILES is a member of a whole family of related formats. The other members are SMARTS (SMiles ARbitrary Target Specification) for description of structural patterns and SMIRKS for description of chemical transformations. Of these, SMILES is by far the most commonly used.

Basic Rules for Writing SMILES

Because of the human-friendliness of SMILES, it is used as an input format in a variety of online databases and other chemistry-related websites. Therefore, it is useful to have at least basic knowledge of its properties and the rules for writing proper SMILES.

As a typical example of linear formats, SMILES does not include molecular geometry and focuses only on molecular topology. By default, SMILES also omits hydrogen atoms; unless specified otherwise, each atom is expected to bear hydrogen atoms to match its most natural valence. When we discard hydrogen atoms, SMILES bears very close resemblance to common linearized formulas as used by chemists around the world, such as CH_3CH_2OH. The corresponding SMILES of ethanol is CCO. However, because SMILES has to be able to express exactly all possible molecular structures, it has much more specific rules than the vague ones of linearized formulas. The following list summarizes some of the basic rules of SMILES notation.

- Atom symbols are written normally; the first letter should be capitalized (usage of lowercase symbols is discussed below).
- Bonds are specified using hyphens (-) for single bonds, equals signs (=) for double bonds, hash marks (#) for triple bonds, and colons (:) for aromatic bonds. Single bonds may be omitted and are implied between two following atom symbols. For example, both C-C=C and CC=C represent propene.
- Atoms that follow immediately after each other (regardless of if a bond is specified or implied) are connected with a bond.

- Branching is achieved using brackets; for example, CCOCC means diethyl ether, whereas CC(O)CC represents butane-2-ol.
- Digits after atom symbols represent additional bonds between atoms with the same digit. This way rings are created. For example, CCCCC means pentane, C1CCCC1 means cyclopentane, and C1CCC1C means methylcyclobutane.
- To simplify encoding of aromatic compounds, special notation was devised. When atom symbols are written in lowercase letters, aromatic bonds are implied between atoms (unless a different bond is explicitly given). For example, benzene can be written as c1ccccc1, whereas while C1CCCCC1 means cyclohexane.
- Atoms without implied hydrogens or with any specific properties, such as charge, atomic mass for isotope specification, and so on, are given in square brackets. Thus, C is methane and [C] is elemental carbon; C[13C](=O)O is acetic acid with ^{13}C atom in the carboxyl group, and CC(=O)[O-] is acetate anion.
- Double bond stereochemistry is encoded using slash (/) and backslash (\) characters to specify relative positioning of substituents around the plane of a double bond in a manner similar to its common depiction in drawings: C\C=C\C means (E)-butane, and C\C=C/C means (Z)-butane.
- Tetrahedral stereochemistry is specified in square brackets, together with other atom properties, using one @ character for anticlockwise and two @ characters (@@) for clockwise layout of atoms around the chiral center. The orientation is given as perceived when looking from the first atom that appears in SMILES in the direction of the chiral center at the other three atoms in the order of the SMILES string. Thus, the symbol used depends on the order in which individual substituents appear in the SMILES string, as demonstrated in Figure 5.2, which shows that (R)-butane-2-ol may be written either as C[C@@H](O)CC or as C[C@H](CC)O.

At the beginning of this chapter, we discussed the importance, advantages, and disadvantages of canonical encodings of chemical structures. Because SMILES was developed with human users as the primary audience, it cannot be canonical by default. However, because of the additional value that canonization brings, SMILES has a specification for a canonical version (Weininger et al. 1989).

Even though InChI (discussed in the next section) is quickly gaining support as the linear format of choice, the fact that SMILES can be read and written by humans

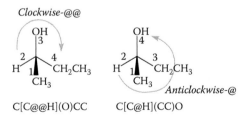

FIGURE 5.2 Specification of tetrahedral stereochemistry in SMILES.

without the need to use a computer, gives it its special place among other common chemical formats and makes it irreplaceable for some types of use.

InChI (International Chemical Identifier)

Basic Characteristics

InChI is a linear format that:

- Has a computer-generated format
- Is not human readable or writable
- Is canonical
- Has unique identifiers for molecules based solely on their structures
- Has moderate, fast growing software support

A format example (InChI of ethanol) is as follows:

```
InChI=1/C2H6O/c1-2-3/h3H,2H2,1H3
```

Format Overview

InChI is a very recent member of the family of chemical formats (McNaught 2006). It is a linear format that was developed in cooperation with NIST and IUPAC and has several important features that distinguish it from other formats. Unlike the other formats discussed above, it cannot describe reactions—only compounds.

The main intention behind the development of InChI was to create a new way of "naming" compounds that would enable computer programs to assign them unique identifiers, regardless of how they are drawn and without the need for a central registration point for such identifiers (as in the case of registry numbers and other similar identifiers). This intent led directly to the fact that InChI cannot be created by humans, because they would not be able to reliably reproduce the steps needed to create such a unique identifier. With this fact in mind, InChI was created to be written and read by computers only. This is in strong contrast to SMILES, which was created specifically to be written and read by humans and even in its canonical form is at least human readable.

The intended use of InChI as an independent unique identifier of compounds directly enforces canonization of the format. Unlike in SMILES, there is no noncanonical form. Also, the canonization of a structure in InChI goes much further than in SMILES, where it is mostly limited to canonical numbering of the atoms and following encoding of the result based on this numbering. In InChI the canonization process also involves very deep normalization of the structure, which transforms several types of drawing conventions (e.g., nitro group with double bonds versus with charged atoms) to one normalized form and detects and normalizes some isomeric forms (tautomers etc.).

Similar to SMILES, InChI does not store atom coordinates. In contrast to SMILES, which by default omits hydrogen atoms that are then added implicitly to match the most common valency of an atom, InChI stores hydrogen atoms but does not store bond orders. These two techniques are just different approaches to the same problem; for a given molecular skeleton, the bond orders and number of hydrogen atoms attached to each heavier atom are interdependent, and one can be used to compute the other. Because both of these formats are linear, they try to optimize for size by removing one piece of the redundant information. The approach taken in SMILES is closer to the praxis of chemists who usually omit hydrogen atoms from drawings, but omitting bond orders used in InChI gets very nicely around the problem of aromatic compounds without the need for a special syntax as used in SMILES.

Another important feature of InChI is its layered structure. Unlike in SMILES, where all data related to one atom are stored in one place, in InChI different properties of the structure are encoded in different parts of the identifier. This organization of the data has one very important advantage: molecules with the same basic structure that differ only in some minor property, such as in stereochemistry or isotopic composition, have the same InChI, with only the exception of the corresponding layer. This makes it possible not only to compare two InChIs to find if they represent exactly the same structure, but to use a more intelligent comparison of two InChI strings to reveal molecules with the same basic structure that differ only in some detail. It is then up to the user to decide which deviations in the InChI are significant for his or her purpose and which are not.

The layer structure of InChI is demonstrated in Figure 5.3, and Figure 5.4 shows several similar structures and the influence different changes in the structure have on the resulting InChI.

The following list briefly mentions all possible InChI layers and discusses some important facts that arise from properties of each layer:

- Main layer
  - Chemical formula: Hill ordered summary (see the discussion of protonation below for examples of unexpected results).
  - Connections: Connectivity of nonhydrogen atoms and assignment of hydrogens to these atoms.
- Charge layer
  - Component charge: Net charge of the molecule (or charges of individual components); no information about position of charge is stored in InChI. It is important to be aware of this; the same InChI will be generated for two structures that differ only in the position of charge.

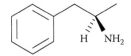

(*S*)-Amphetamine

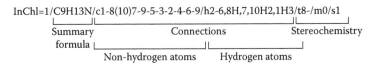

FIGURE 5.3 Layer structure of InChI.

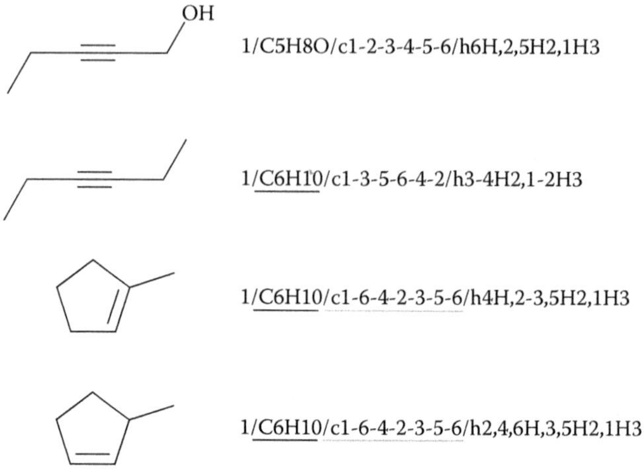

1/C5H8O/c1-2-3-4-5-6/h6H,2,5H2,1H3

1/C6H10/c1-3-5-6-4-2/h3-4H2,1-2H3

1/C6H10/c1-6-4-2-3-5-6/h4H,2-3,5H2,1H3

1/C6H10/c1-6-4-2-3-5-6/h2,4,6H,3,5H2,1H3

FIGURE 5.4 Influence of chemical structure on individual InChI layers.

- Protons: Protonations (and deprotonations) are treated separately from other types of charge changes. This makes InChI of acetic acid (InChI=1/C2H4O2/c1-2(3)4/h1H3,(H,3,4)) and acetate anion (InChI=1/C2H4O2/c1-2(3)4/h1H3,(H,3,4)/p-1) identical in all layers but this one. The unexpected result of this feature is that the summary formula does not correspond to the described structure in such cases; acetate anion is $C_2H_3O_2$, not $C_2H_4O_2$. Like the component charge layer, no information about the position of protonation is given.
- Stereochemical layer
  - Double bond and sp2 stereochemistry
  - Tetrahedral (sp3) stereochemistry
- Isotopic layer
- Fixed-H layer: This layer is optional, and its inclusion into InChI may be switched on or off in the InChI software. By design, structures differing only in position of movable hydrogens are encoded with one InChI. This layer makes it possible to distinguish between these forms as demonstrated in Figure 5.5.

Even though InChI is a relatively new format (the first stable version of the InChI software was published in 2005), it has already gained broad acceptance in the chem-informatics world. It is supported by most of the major chemical drawing programs and is used in such databases as PubChem (http://pubchem.ncbi.nlm.nih.gov/) and NIST Webbook (Linstrom and Mallard 2005).

The features of InChI make it usable as a unique identifier of a molecular structure as well as a format for data storage (with limitations implied by its design, such as the absence of atom coordinates and charge localization, or delocalization of hydrogen atoms, which makes it impossible to distinguish between individual tautomeric forms without the presence of the fixed hydrogen layer).

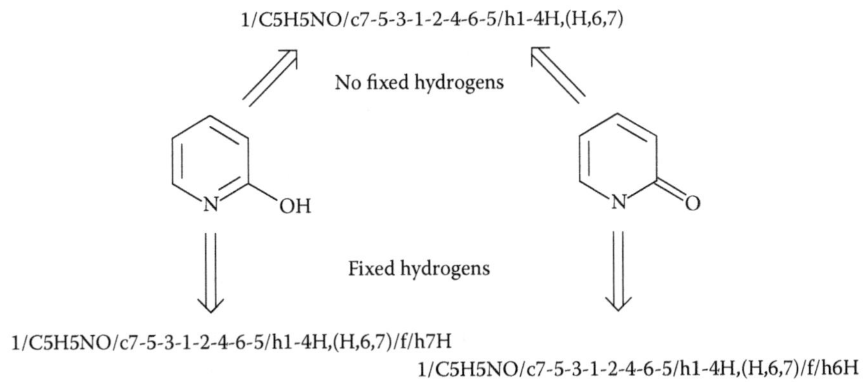

1/C5H5NO/c7-5-3-1-2-4-6-5/h1-4H,(H,6,7)

No fixed hydrogens

Fixed hydrogens

1/C5H5NO/c7-5-3-1-2-4-6-5/h1-4H,(H,6,7)/f/h7H

1/C5H5NO/c7-5-3-1-2-4-6-5/h1-4H,(H,6,7)/f/h6H

FIGURE 5.5   Demonstration of the fixed hydrogen layer influence on the final InChI.

In the following section we discuss InChIKey, a format derived from InChI, which may replace InChI in cases where a simple unique identifier of a compound is needed.

INCHIKEY

Basic Characteristics

InChIKey is a linear format that:

- Is a computer-generated format based on InChI
- Serves as a "fingerprint" of a molecular structure
- Has a fixed length
- Is not convertible back to the original structure
- Has low software support that is expected to grow fast (it's a very new format)

A format example (InChIKey of ethanol) is as follows:

LFQSCWFLJHTTHZ-UHFFFAOYAB

Format Overview

The most important property of InChI is that it provides a unique identifier for a compound based solely on its structure. This feature would seem to make it ideal for purposes like database indexing, publication inside articles as a form of alternative chemical naming, online display, searching of chemical compounds, and so on. Even though InChI really is usable for the mentioned purposes, it has several features that make it less then ideal for these tasks. The main weakness of InChI in these cases is its length, which is proportional to the size of the encoded molecule. While this feature is natural and unavoidable for a complete format, it means that the larger the molecule is, the more space we have to reserve for its representation, either in a database, in a printed publication, or on a website. Also, the content of the identifier does not form one homogeneous string, which means it will be split into smaller parts by search engines or in text processing programs that are designed to work with common text and are not aware of the notation of InChI. In this respect, registry numbers, such as CAS RN, have a big advantage of being short, usually of well-defined structure and limited length. However, such identifiers are neither deducible from the structure nor convertible back to the structure without the need of database lookup.

The idea behind InChIKey is to bring the advantages of a fixed-length registry number together with the computability of InChI. The decoding of the identifier had to be sacrificed to its fixed length.

Fortunately, in computer science there already exists a well-known and widely used solution to the problem of computation of a fixed length fingerprint from a variable arbitrary sequence of bytes; it is called hashing. This technique is used in many ways (for password storage, file content checks, digital signatures, etc.), and there are several strong and well-tested hash algorithms to choose from. The hash algorithm used in InChIKey is SHA-256 truncated to match the length selected for InChIKey.

The InChIKey has a fixed length of 25 characters, divided into two groups (14 and 10 characters long) by a hyphen. Both parts of the InChIKey include encoded hashes of different layers of the original InChI string. The second part also contains a checksum character (the last one), which enables checking of validity of the whole InChIKey after transcription or other transfer. Such a two-part arrangement of the InChIKey structure was selected to at least partly preserve the layered structure of InChI. Therefore, the most important data about a structure are contained in the first part, and the second part is reserved only for information about isotopes, stereochemistry, and fixed hydrogens. This fact means that, for example, stereoisomers will differ only in the second part of the identifier.

InChIKey was carefully designed with online use and search engines in mind. It uses only uppercase letters, so there are no characters that would be considered as word-splitting by a search engine. This means the InChIKey should be processed as two words and be easily searchable on the Internet. Because only uppercase letters are used, false collisions based on case insensitivity of the search are avoided. It is a very compact form, and the presence of an internal validation mechanism makes InChIKey very resistant against possible errors introduced during transfer, such as wrong transcription, line-wrapping in an email client, and so on, which makes it in this sense superior to both InChI and SMILES.

Because of the nature of InChIKey (and every hash in general), collisions are possible. This fact comes directly from the limited number of possibilities a 25-characters-long string can contain. Even though collisions of InChIKeys are inevitable in the future, it is not possible to say when the first collision will occur. The official InChI documentation (documentation published with the InChI source code, version 1.02-Beta; http://www.iupac.org/inchi/download/index.html) states that the probability of a collision in a set of 1 billion InChIKeys is $2.0 \times 10^{-20}\%$. However because the second part of the InChIKey is based on InChI layers that do not exist (are empty) for many structures (such as isotopic layer, stereochemistry layer, etc.), a more realistic estimate must be based on collisions in the first part of the InChIKey alone. In this case the same source states that the probability of a collision in a set of 1 billion structures increases to $2.7 \times 10^{-9}\%$. However, even this means that unless we are extremely unlucky, InChIKey should remain unique for quite a long time. It was tested against a set of 77 million structures (both existing and generated) and no collision was found.

The hash origin of InChIKey also means that it is not convertible back to the original InChI or molecular structure, because for each InChIKey there is an unlimited number of possible matching input values. Although this might seem to be a drawback of the format, it is simply the price of the fixed length of the identifier. When a readable identifier with no possible collisions is needed, InChI (or canonical SMILES) should be used.

At the time of writing of this chapter, InChIKey is a very new format. However, once it starts to gain support in chemical software and online databases, it will have the potential to replace conventional registry numbers and provide a standardized, easy to use chemical identifier that is based solely on the molecular structure.

OTHER FORMATS

In this section we will briefly discuss formats that were not mentioned previously but might be interesting to the reader, either because they offer some specific features or because they are used in specific fields.

Formats Used in the Protein Data Bank

The Protein Data Bank (PDB; http://www.pdb.org) is the worldwide repository of three-dimensional structural data of biological macromolecules, such as proteins and nucleic acids (Berman et al. 2003). The Protein Data Bank uses several text file–based formats for data deposition, processing, and archiving. The oldest of these is the Protein Data Bank format (Bernstein 1977), which is used both for deposition and for retrieval of results. It is a plain-text format whose main part, a so-called primary structure section, contains the atomic coordinates within the sequence of residues (e.g., nucleotides or amino acids) in each chain of the macromolecule. Embedded in these records are chain identifiers and sequence numbers that allow other records to reference parts of the sequence. Apart from structural data, the PDB format also allows for storing of various metadata such as bibliographic data, experimental conditions, additional stereochemistry information, and so on. However, the amount of metadata types available is rather limited owing to the age of the PDB format and to its relatively strict syntax rules.

The PDB format was created in the 1970s and has gained much software support since that time. The format is still being developed and new revisions of the format are published. Even though it is not as widely used in the area of general chemistry, it certainly deserves attention as the major format in the field of structural biology.

Although the PDB format has served the community well since its inception, it is widely recognized that the current PDB format cannot adequately express the large amount of data (content) associated with a single macromolecular structure and the experiment from which it was derived in a way (context) that is consistent and permits direct comparison with other structure entries. Therefore, an alternative approach has been developed under the auspices of the International Union of Crystallography (IUCr), leading to another plain-text format: the mmCIF format (macromolecular Crystallographic Information File; Bourne et al. 1997). The mmCIF format represents an extension of the Crystallographic Information File (CIF) used for describing structures of small molecules. Compared to the PDB format, mmCIF is more flexible, allowing not only for storing data in predefined categories, but also for defining of users' own categories. Also, the mmCIF format is more suitable for automated computer processing. Therefore, it is used internally in PDB as its primary storage flat file format. The price for its flexibility is its rather complex structure. Thus, among experimentally oriented structural biologists, the PDB format still prevails and won't be replaced by the mmCIF format in the near future.

In accordance with the general trend in informatics to base file formats on XML, PDB has also created its newest file format in this way. The result is called PDBML (Westbrook et al. 2005), and even though the name would suggest some relationship to the old PDB format, it closely resembles the mmCIF format. Both mmCIF and PDBML formats are used internally in PDB, and their more complex nature will probably never make them as widely used as their older sibling. However, the importance of the PDB itself makes any format it uses worth knowing.

SYBYL Line Notation (SLN)

SYBYL Line Notation (Ash et al. 1997) is a linear format that to some extent resembles SMILES. The most important feature that these two formats have in common is human readability. However, the authors of SLN have made several design decisions that make it different from SMILES in several important points:

- Hydrogens must be given as part of the formula; no hydrogens are implied based on standard valences of atoms.
- Aromatic bonds are given explicitly using a colon (:); it is not possible to use a lowercase letter for an atom symbol for specification of aromatic bonds.
- Stereochemistry is handled differently and can be used to express relative stereochemistry, mixtures of enantiomers, and so on.
- SLN includes support for substructure search queries. In the SMILES world, this task is handled by other derived formats.
- Additional information about an atom, such as charge, is given in square brackets after the atom symbol, not together with the atom symbol. The types of additional information are not limited and may be even user defined.
- SLN is able to store atomic coordinates.
- Rings are encoded using a slightly different method that uses atom number references.

Even though SLN does not have the widespread support that SMILES has, it is used in several systems, and some of its properties mentioned above make it more advanced and general than SMILES.

CONVERSION OF CHEMICAL FORMATS

In this section, we will very briefly discuss the topic of conversion between different chemical formats. Even though every chemical editor can export its drawings into several external chemical formats and tools are available for batch conversion between different formats (such as the widely used open-source program Open

Babel [http://openbabel.sourceforge.net/]), it is important to keep in mind the limitations of each format. For example, conversion from SMILES, which does not store atom coordinates, into Molfile would result either in the molecule having undefined geometry or the geometry being supplied by the conversion tool, which might be completely different from the real one.

It is even more important to keep such limitations in mind when one format is used as an intermediary because the conversion is not possible in a direct way. Choosing a linear format for this task would inadvertently lead to loss of information about geometry of the converted structure, even though both the original and the final format it.

Keeping such limitations in mind, it might be said that for the widely supported formats discussed in the above sections, there is no problem converting data between any two of these, for example, using Open Babel.

ONLINE PUBLICATION OF CHEMICAL DATA

In this section, we will discuss which formats to use for online publication of chemical data. Because different publishers of data have different goals and demands in this area, we will concern ourselves with several typical scenarios of publication of chemical structures.

PUBLISHING DATA FOR SEARCH ENGINES

The goal here is to make chemical structures available for search engines. Typical deployment includes journal publishers, vendors of chemicals, individual scientists, and working groups.

On today's Internet it is vitally important for a website to be visible in a search engine. In the following paragraphs we will discuss how to make chemical structures available to search engines so that the user can easily find them. Although we will discuss only publication of structural data, it is important to mention that even with the development of new linear formats such as InChI and InChIKey, which may drive increased usage of general search engines for structural searches, most users would use chemical names to search for information about a compound. Therefore, whenever possible, it is very important to include chemical names of published structures.

When publishing chemical structures for search engines, all file-based formats are usually out of the question. The reason is simple: Search engines work on the scale of characters and words and are not prepared to compare whole files. Thus, we remain with the selection of available widespread linear formats: SMILES, InChI, and InChIKey.

The most important factor that determines the possibility of using a format for online searching is the existence of a canonical form of the format. Without canonical encoding, there are simply too many possibilities of search input (even just the most probable ones) that would have to be tried.

Both InChI and InChIKey are canonical by default, and there is no other option for them. The situation is different for SMILES, where both a canonical and a noncanonical form exists. For this reason InChI (or InChIKey) is probably better, because the canonization must be done both on the side of the publisher and on the side of the user who performs the search. Because the latter is out of our control, it is safer to use a format that does not allow noncanonical forms. Also, the canonization algorithm used in InChI has stronger normalization, thus making the search even more efficient.

One other thing we have to consider when selecting the right format for our purpose is the compactness of the format. The problem here is that search engines are usually word based and ignore some characters, such as punctuation. Therefore, it is advantageous to have a format that does not "fall apart" when read by a search engine (e.g., SMILES for biphenyl c1ccccc1-c1ccccc1 will be almost surely read as two separate words," and it would thus be impossible to effectively search for the biphenyl SMILES without many false hits where benzene is found). In this respect InChIKey is by far the best, because it is formed by two compact parts that contain only letters; thus, both parts will be treated as one word by search engines. Because the first part of InChIKey contains most of the structural information, it is in this case advantageous that the two parts will be searched separately. The other two formats, InChI and SMILES are often much longer than InChIKey and contain many characters that are treated as word splitting. SMILES has one additional problem: Because search engines usually do not distinguish between lowercase and uppercase letters, it is impossible to distinguish between benzene c1ccccc1 and cyclohexane C1CCCCC1.

Lastly, we should take into account the probability that the user will use our particular format for searching. This fact depends not only on how well established the format is, in which SMILES would probably win and InChIKey would be an outsider, but also on the user's perception of which format is suitable for this task. Because InChI, and InChIKey even more, were marketed from the beginning as suitable for online searches, it is very likely that the user will use one of these formats for searching, rather than SMILES, which does not have such an association. Of course, we could solve this problem by simply publishing our data in all three formats. The decision should be made based on our resources, but it is in general very easy to convert between the formats automatically.

Taking all of the above into account, probably the best format for search engines is InChIKey. Its only disadvantage at this time is its lower penetration into the media, but we can expect it to gain popularity very quickly, especially in free online sources.

We should briefly mention the possibilities of inclusion of the structural data into a web page. In many cases we would just place it somewhere in the text, below a picture that represents the structure of the corresponding molecule or in a separate table. However, it might not suit our purpose to place some strange and possibly several-lines-long identifiers into our text. In this case, it is possible to make the identifiers part of the page but use styling to hide the information from the user and still make it available to the search engine robot. We have successfully used this approach to invisibly add InChI into the online version of the IUPAC Compendium of Chemical Terminology (http://goldbook.iupac.org; Nic et al. 2006).

Publishing Data for Online Display

The goal here is to display chemical structures online on a web page. Typical deployment includes educational websites and online databases.

Direct online display of common chemical formats is not possible without use of specialized programs. Because of this fact, it is often advisable to convert them into a web-friendly graphical format, such as PNG (Portable Network Graphics) or GIF (Graphics Interchange Format). Although JPEG (Joint Photographic Experts Group) is also usable, its nature leads to development of undesired artifacts around sharp edges of text and lines. For this reason lossless formats, such as PNG, are more suitable. In specific cases or as an addition to a bitmap version of a drawing, a vector version, such as SVG (Scalable Vector Graphics) or EPS (Encapsulated PostScript), may be used. The advantage is much better printout quality when such a format is used.

Sometimes, though, it is possible to use conversion to graphic formats to display all properties of a structure, especially three-dimensional structures. Even though it is possible to create two-dimensional renderings of such structures and even provide a stereo-view, three-dimensional structures are best visualized by direct manipulation, such as rotation and magnification. These features cannot be provided by simple bitmap graphics.

In such cases, it is possible to either use specialized software, designed for the particular purpose of chemical data display, such as Jmol (http://jmol.sourceforge.net/) or Chime (http://www.mdl.com/products/framework/chime/), or to convert data into a three-dimensional-friendly format, such as VRML (Virtual Reality Modeling Language).

The former approach gives the user an environment specializing in chemistry, with options and tools not found in general software (such as the ability to choose between ball and stick, wire-frame, and other representations of a molecule). The programs are usually able to work with many input formats, such as Molfile or CML, and there is therefore often no need to convert the published data. Of the tools mentioned above, Jmol deserves special attention because it is an open-source program that works as a Java applet inside the browser, so the user does not need to install any special tools (of course, Java is mandatory).

Conversion to a three-dimensional format such as VRML has its advantages if the audience can be expected to have some experience with the software or the author has special needs not met by the above-mentioned chemistry software.

PUBLISHING DATA FOR SHARING

The goal here is to provide data in a way that makes them easy to download, possibly all at once. Typical deployment includes scientific organizations and free online databases.

There are several examples on the Internet of websites that are actively working toward providing free and unrestricted access to scientific data for the scientific community and general public. One of the biggest examples that is also closely related to chemistry is PubChem (http://pubchem.ncbi.nlm.nih.gov/), an online database of chemical compounds, their structures, and chemical and other related data. PubChem not only provides its data online on its webpages, but it also makes the chemical content easily accessible using automated tools (in this case, compressed SDfiles containing tens of thousands of structures each are used).

When publishing data for others to use, be it on a multimillion or a few dozen scale, it is important to evaluate which format will preserve the most of your data and still be as widely supported as possible. Although it is possible to simply put online the files that are natively used by your chemical editor, thus being sure that even coloring of bonds and fonts are preserved, you will inadvertently limit the number of users who will be able to use your data. However, using a linear format might lead to loss of information in cases where geometry matters, such as in results of theoretical calculations, X-ray measurements, and so on. For this reason it is advisable to use one of the widely adopted file-based chemical formats: CTFile formats (such as SDfile) or CML.

Of course, when data are available in one of the above-mentioned formats, it is very easy to automatically convert them to other formats such as InChI or InChIKey to make the most of them and prepare the website for search engines.

STORING CHEMICAL DATA

In this section, we will briefly discuss the advantages and disadvantages of the above-described formats when it comes to storing chemical structures for personal, laboratory, or similar use. We assume that any large chemical database project will be based on a ready-to-use commercial solution and will therefore focus on small-scale or occasional users willing to preserve their data and make them searchable. Our goal will not be to create a fully working system for chemical structure storage, but rather to point out possible solutions and warn against common mistakes.

DATA STORAGE

When storing chemical drawings, we have to ask ourselves what kind of data we actually have. Are we dealing purely with well-defined chemical structures, or do we also need to store structures with variable substitution, additional text, or even related graphics? The reason for this question is that the formats discussed in the previous sections are all designed only for the purpose of storing chemical structures and cannot reliably describe all additional features of the drawing. Therefore, the following paragraphs pertain only to storage of "pure" chemical structures.

However, even if we work only with well-defined structures, it is always a good idea to store the source files (usually in the native format of the software that was used to create them), because in this way we can make sure no part of the drawing is lost in the conversion. However, native formats usually have the disadvantage of being bound to one specific program, and options for their automatic processing are limited. For this reason native data should be auxiliary, and the data for everyday use should be stored in some widely supported, well-documented format, probably one of those described above. This approach would also create a future-proof copy of our data, because proprietary formats tend to be problematic to use after 10 years or more.

Linear formats are usually not good candidates for primary data storage, simply because they do not contain the molecular geometry. Therefore, it is advisable to use one of the file-based formats for this purpose—either Molfile or CML. These are very similar in their capabilities and have good support in various chemical software packages, with Molfile usually having slightly smaller files and even wider support than CML.

However, file-based formats are not directly comparable when a search in a database is performed and special chemistry-aware tools have to be used. For the purpose of searching in the database, linear formats might be more useful. This topic is discussed in the following subsection.

Data Indexing

Data storage is only one aspect of the problem of chemical structure handling. The other, at least as important, part is searching and, more importantly, finding of the right data. To perform a full-featured search in a chemical database, as we know from SciFinder or CrossFire, we need specialized software (and powerful hardware if we have larger amounts of data). However, in a personal catalog used in everyday laboratory praxis, people most frequently need to simply find if a compound is somewhere in their cupboard or if someone from their group has its spectra measured. For the purpose of finding if a structure is or is not present in a database, any unique identifier of the structure will do. If we discard the possibility of using some already used registration number, for which we would need to first look up the appropriate number, we have already described three alternatives that provide exactly this: canonical SMILES, InChI, and InChIKey. Each of these has advantages and disadvantages. All of them are computer generated based on a drawn structure and can be easily generated from either Molfiles or CML (which we already decided to use as primary formats for our data).

SMILES has the additional feature of being human readable, but this is not very important in our model case. InChI, and InChIKey by inheritance, features a much better and robust normalization of structures; for example, two different tautomeric forms have the same InChI, but different canonical SMILES. Also, the layered structure of InChI gives us the possibility of excluding some particularity of a structure, such as its stereochemistry, from the search when needed. This is not possible using InChIKey or SMILES.

The main advantage of InChIKey in this mode of use is its fixed length and short, compact form. These features reduce the chance of possible errors being introduced when the identifier is copied from an email or written by hand, which might be considerable for larger structures for which InChI can easily span several lines of text. The simple format of InChIKey makes it even possible to use it as part of a directory or file name, thus creating a very simple database without the need for database software. However, there is always the possibility of InChIKey collision, even though the chance is miniscule. Therefore, cautious users should augment it with some other kind of identifier, just in case a collision occurs.

Summary

The following list summarizes a possible setup of formats for a reliable, redundant chemical structure catalog based on the recommendations above:

- InChIKey as the primary key for quick lookup into the database
- InChI as auxiliary information to InChIKey (for the purpose of readability and possible layer-by-layer comparison and as a safety precaution in case InChIKey collision occurs)
- Molfile or CML as the primary source of structural data that can be used to create the above-mentioned identifiers
- Original native file as a safety precaution and for presentation purposes

CONCLUSION

In this chapter we have described the most commonly used formats for chemical structure storage. We have discussed in detail their properties and suitability for various scenarios of online publication and data storage.

REFERENCES

Ash, S., M.A. Cline, R. Webster Homer, T. Hurst, and G.B. Smith. 1997. SYBYL Line Notation (SLN): a versatile language for chemical structure representation. *J. Chem. Inf. Comput. Sci.* 37:71–79.

Berman, H.M., K. Henrick, and H. Nakamura. 2003. Announcing the worldwide Protein Data Bank. *Nat. Struct. Biol.* 10:980.

Bernstein, F.C., T.F. Koetzle, G.J.B. Williams, E.F. Meyer, Jr., M.D. Brice, J.R. Rodgers, O. Kennard, T. Shimanouchi, and M. Tasumi. 1977. Protein Data Bank: a computer-based archival file for macromolecular structures. *J. Mol. Biol.* 112:535–542.

Bourne, P.E., H.M. Berman, K. Watenpaugh, J.D. Westbrook, and P.M.D. Fitzgerald. 1997. The macromolecular Crystallographic Information File (mmCIF). *Meth. Enzymol.* 277:571–590.

Chartrand, G. 1984. *Introductory Graph Theory.* New York: Dover.

Dalby, A. et al. 1992. Description of several chemical-structure file formats used by computer-programs developed at Molecular Design Limited. *J. Chem. Inf. Comp. Sci.* 32:244–255.

Holliday, G.L., P. Murray-Rust, and H.S. Rzepa. 2006. Chemical markup, XML and the Worldwide Web. 6. CMLReact; An XML vocabulary for chemical reactions. *J. Chem. Inf. Mod.* 46:145–147.

Kuhn, S., T. Helmus, R.J. Lancashire, P. Murray-Rust, H.S. Rzepa, C. Steinbeck, and E.L. Willighagen. 2007. Chemical markup, XML, and the World Wide Web. 7. CMLSpect, an XML vocabulary for spectral data. *J. Chem. Inf. Mod.* 47:2015–2034.

Linstrom, P.J., and W.G. Mallard (Eds.). 2005. *NIST Chemistry WebBook, NIST Standard Reference Database Number 69.* http://webbook.nist.gov/.

McNaught, A. The IUPAC International Chemical Identifier: InChI: a new standard for molecular informatics, *Chem. Int.* 28(6):12–14.

Murray-Rust, P., and H.S. Rzepa. 1999. Chemical markup language and XML. 1. Basic principles, *J. Chem. Inf. Comp. Sci.* 39:928–942.

Murray-Rust, P., and H.S. Rzepa. 2003. Chemical markup, XML and the Worldwide Web. 4. CML schema. *J. Chem. Inf. Comp. Sci.* 43:757–772.

Nic, M., J. Jirat, and B. Kosata. 2006. Chemical terminology at your fingertips. *Chem. Int.* 28(6):28–29.

Trinajstic, N. 1992. *Chemical Graph Theory*, 2nd ed. Boca Raton, FL: CRC Press.

Trudeau, R.J. 1993. *Introduction to Graph Theory.* New York: Dover.

Weininger, D. 1988. SMILES: a chemical language and information system. 1. Introduction to methodology and encoding rules. *J. Chem. Inf. Comp. Sci.* 28:31–36.

Weininger, D., A. Weininger, and J.L. Weininger. 1989. SMILES. 2. Algorithm for generation of unique SMILES notation. *J. Chem. Inf. Comp. Sci.* 29:97–101.

Westbrook, J., N. Ito, H. Nakamura, K. Henrick, and H.M. Berman. 2005. PDBML: the representation of archival macromolecular structure data in XML. *Bioinformatics* 21:988–992.

6 Chemical XML Formatting

Miloslav Nic

CONTENTS

INTRODUCTION

Data mining applications aim at the automatic discovery of new information in available data. Because extraction of data from unstructured text is very difficult, current applications usually work with data that is in some way structured.

In recent years, XML (eXtensible Markup Language) has become the technology of choice. It is *extensible* because it allows users to represent or define their own elements while facilitating the exchange of semistructured data across different information systems with different user-defined elements. It is not surprising that nowadays data are commonly stored in XML and that non-XML data are routinely converted to an XML format before further processing. Therefore, anybody interested in data mining applications should understand the fundamental aspects of XML technologies in order to follow modern trends and communicate with experts in data mining.

This chapter provides broad overview of relevant aspects of XML technologies and discusses XML usage in chemistry and other relevant fields.

XML TECHNOLOGIES

XML has become very popular in recent years. It is difficult to find an area of computing that does not use XML in some way. Software applications nowadays work with XML data, programming languages offer libraries specializing in XML programming, and thousands of developers have gained experience with XML usage. Considering this phenomenal success, it is highly improbable that XML technologies will fade into oblivion in the near future. Neither material nor human investments in this area are therefore endangered by the potential disappearance of the underlying technology.

Because many diverse communities with different professional languages have adopted XML, terminology misunderstandings are common. Before focusing on the details of XML, it is important to define basic terms and their mutual relations: XML documents are collections of letters, digits, and other characters that conform to a set of rules. Such documents can reside in files or databases or be transiently generated by a program and immediately processed further without leaving computer memory.

XML syntax provides rules to which every document must conform. These rules specify which characters are permitted inside the document, how they can be ordered, and how this ordering provides structure to information present inside document. If there is just a single violation of these rules, XML software—any software that can process XML documents in a way specified by these rules—is required to stop processing and quit.

When a document is parsed (read) by a software application, the contained data is transformed into structures that can be further processed by computer. This resembles the process of human reading, where the eyes see long rows of characters while the brain gives meaning to these rows. People are very good in discovering structures hidden inside texts, but computers need much more help.

Markup languages like XML provide formal ways how to present structured information to computers. They augment unstructured text with structural and logical hints that simplify programming tasks to a manageable level. Many markup systems exist to augment the text with these hints, but XML syntax provides some clear advantages over these alternative systems. XML is a tool of choice for the markup of syntactically rich structures.

XML namespaces provide a mechanism for automatic recognition of markup languages. Several markup languages can be used in a single XML document without a dramatic increase in the complexity of processing software.

XML SYNTAX

The basic rules of XML syntax are straightforward and can be learned in a few hours. Some problematic areas require a much more thorough understanding, but such cases are usually not encountered in common practice. The following text provides a short overview and the rationale behind the syntactic rules. Interested readers will easily find many materials of different complexity on the Internet.

The syntactic rules are specified in the World Wide Web Consortium (W3C) standard, "Extensible Markup Language (XML)." The first edition was issued in 1998 (Bray et al. 1998); the currently valid forth edition was published in 2006 (Bray et al. 2006b). If a document conforms to all rules given in this specification, it is said to be "well formed."

The basic structural units of XML documents are elements. With the help of elements, any document can be separated into parts that are logically connected. So if an XML document contains a list of molecules, each molecule can be completely described inside its own element without any reference to the rest of the text. This ensures that each element is safely and independently processed.

Molecules consist of atoms. The atoms represent a different logical entity from molecules, so it is natural to express their presence with the help of different elements. At the same time, it is necessary to express the notion that a molecule is a collection of several atoms. In XML, some elements can be children of other elements, and with this hierarchical construction, the concept of a molecule consisting of a set of atoms is naturally expressed. Similarly, atom elements can contain children elements describing properties such as symbol, charge, and so on.

Although these hierarchic structures can be easily expressed with drawings, it is not easy to express them with plain text. Mathematicians encounter similar problems when writing mathematical expressions where ordering of different operations must be performed. Parentheses were invented to solve these problems, as they provide syntax to show the intended grouping of expression parts.

Whereas XML syntax is based on similar principles, usage of parentheses would not be practical. An XML document can be many thousands of pages long, and finding a matching parenthesis would be a nontrivial task. Thus, it is important to have some syntax for naming of the parentheses so that correct pairing of their starts and endings can be easily checked. In XML, such named parentheses are called tags; start-tags open and end-tags close logical parentheses.

Because the parentheses are commonly used in many different texts, pointed brackets (< >) are used instead. The start-tags are marked as "<molecule>." The end-tags are similar, but they contain a slash character (/) before the name (e.g.,

</molecule>). Slash characters distinguish start-tags from the end-tags. If there are no elements or other text between start and end tags, the end tag can be replaced by a slash preceding the closing > of the start element, as in "<electron/>."

Although it is not immediately apparent from mathematical notation, the parentheses never overlap. The inner parentheses are always closed before the outer ones; otherwise, their meaning would not be unequivocal. According to the same logic, if a start-tag of one element is followed by the start-tag of another element, the end-tag of the second element must precede the end-tag of the first one. In short, elements in XML never overlap, a child element is always fully enclosed by its parent element. Thanks to this rule, a hierarchical structure can always be unequivocally understood from XML notation.

The hierarchical structure provides a unifying organizational theme inside a document, but if several elements were without parents, it would be difficult to specify their mutual relations. Another rule was therefore added: In a well-formed document only one element is without a parent. Such an element is called the root element, and all other elements are either its children or descendants. Although the start-tag of this element is therefore the first text in the document, and its end-element is the last text, special types of text called comments and processing instructions can appear anywhere. Comments are ignored by XML applications. Processing instructions are intended for processing software and do not provide any document data in a well-designed language.

The range of characters that can be used for tag names is restricted. For common usage it suffices to remember that tag names can start only with letters or an underscore character (_), followed by an unlimited number of other letters, digits, and characters: _, -, ., and : (with special meaning discussed below). The letters and digits are not restricted to the English alphabet; they can come from virtually any world language. The default encoding of XML documents is Unicode, which can express characters in virtually all existing languages.

Although any information can be expressed by elements, sometimes a simpler notation would be sufficient. In XML, information without further internal structure can be provided inside start-tags as an attribute. The attribute syntax starts with its name and is restricted by the same rules as element names followed by an equal sign (=). The value of an attribute is given in text delimited by apostrophes ('...') or quotes ("..."). Names of elements are separated from attributes by white-space characters (spaces, tabulators, new lines). The same element is prohibited from having several attributes of the same name (<molecule name='benzene' id="i4"/>).

Element and attribute names starting with "XML," "xml," or any other case combination are reserved for possible further usage. It is called a well formedness error if an element name starting with "xml" is used in a non-W3C defined name. For performance reasons, some XML parsers do not check for this constraint, but nevertheless it is very imprudent to use such names. When searching for names of containers for some XML data, it is tempting to use such names, so this error is surprisingly common.

When a parser reads an XML document and encounters a < character, it automatically assumes that it encountered either a start- or end-tag. This rule simplifies the writing of XML reading programs and makes them very fast, but it also causes a serious complication: It makes inputting expressions like "1 < 3" nontrivial. Values of attributes are surrounded either by apostrophes or quotes, so combined usage of both apostrophes and quotes causes similar problems in attribute values. Prohibition of such characters in XML texts is not an option. A special notation therefore had to be introduced that enables an alternative specification of characters. If the < character is used in text, it must be written as "<". Similarly, quotes and apostrophes can be replaced by """ and "'" respectively.

XML NAMESPACES

Elements and attributes provide structure to the document data, allowing different parts of the document to be processed by different programs. Required software may be selected manually, but the selection can be automatized by systemic usage of file suffixes or other hints. Unfortunately, an approach in which information about the markup language is external to the XML document is fragile. The files can be renamed or directories moved, and external information is lost forever.

A more robust system would incorporate markup language information inside the XML documents for selection purposes. The name of a root element is a convenient selection criterion, but only if every markup language uses unique root elements. If there were only several languages around, it would not be a difficult problem, but given the diversity of the Internet, such uniqueness is impossible to achieve.

The XML namespaces provide a mechanism for generation of unique names for each markup language without unduly restricting the choice of available names. The XML name of each element and attribute consists of two parts: the namespace part and the local part. The same local names can be used in many different markup languages, as the namespace part provides a means for disambiguation. If the namespace is truly unique worldwide, then there is no possibility of a name clash, as the name of an element is given by the combination of both parts. With such mechanism in hand, it is necessary to find a process that would guarantee worldwide uniqueness of namespaces. A global repository of namespaces would provide a solution but unfortunately a very impractical one. The Internet makes such a repository feasible, but the nightmare of setting up and running the database in the jungle of financial and political considerations would result with high probability in disaster.

Fortunately, a centrally managed system is not the only solution. From a practical point of view, it is sufficient to adopt some empirical rules that make creation of duplicated namespaces highly improbable. The development of the Internet brought about such a device in the shape of unique human-readable addresses: URLs. Although namespace names are not required to have any specific form, it has become a common practice to create new namespaces that resemble absolute URL addresses. Both organizations and individuals with a need for their own namespace usually own a domain name. Because domain names are unique, if a namespace starts a domain name, then the probability of ambiguity is miniscule.

Usage of URLs causes some problems that do not stem from technical limitations but from human misunderstandings. Inexperienced users often believe that if something looks like a URL, it needs to be connected with the Internet. They express concerns about the necessity of Internet connection for XML processing and are surprised that attempts to visit a page with addresses given in namespaces often fail.

In fact, any namespace string is just an identifier. "http://zvon.org/my/namespace" and "This-is-my-namespace" are equally valid namespaces; the first is just visually similar to an Internet address. XML documents containing namespaces can be used without the Internet, and URL addresses that lead to the same location on the Internet would represent two different namespaces if they differ by just a single character.

An XML namespace standard (Bray et al. 2006a) specifies the rules for proper namespace usage. These rules are not complicated, but they are often misunderstood, so erroneous use of namespaces is quite common.

Documents that use namespaces can be recognized by the presence of colons (:) in element or attribute names. The characters after the colon give the local name, whereas the characters before the colon are reserved for namespaces, as in "html:body." Because namespace names can be very long and may contain characters prohibited in XML names, the string preceding the colon is not directly a namespace but a prefix identifier that is mapped to the particular namespace elsewhere in the XML document.

Prefixes can be viewed as mathematical variables that represent real values. This means elements "<myNamespace:book>" and "<myNamespace:book>" can be different and "<htm:head>" and "<xhtml:head>" can represent the same elements. It depends on the assignment of the namespace to the prefix, and this is done by namespace declarations. Namespace declarations resemble attributes, but their names start with "xmlns" followed by a colon and the prefix name. The following value in quotes specifies a namespace. So if the element "html:body" is used, the declaration xmlns:html="http://www.w3.org/1999/xhtml" must be provided if the body element comes from the XHTML specification. The number of namespace declarations in a single document is not restricted, so several markup languages can be used in one XML document without danger of naming collisions. Unfortunately, not every important XML markup language defines and uses its own namespace. If there is no namespace available, it is necessary to recognize the markup language by other means and in the context of data mining, where a huge collection of documents from many different sources may be processed, this becomes a challenging problem.

VALIDATION OF XML DOCUMENTS

The well-formedness requirements remove many problems that plague processing of other types of formats, so parsing of XML documents is relatively simple, but data processing does not end with the reading of data into computer memory. It includes other programming tasks. Any experienced programmer will attest to the fact that taking care of all possible errors in the input data can be time consuming, and sometimes the programming code addressing incomplete or erroneous data is several times longer than the processing one.

All XML documents consist of elements and attributes, so they look very similar. It is therefore not surprising that some sorts of errors, such as missing attributes and the wrong order of children elements, reoccur in many different usage scenarios. This regularity of errors opens a way for simpler discovery of common problems with the help of general software tools.

These general tools are called schemas, and the process of checking XML documents against schemas is called validation. A valid XML document is a document that conforms to some schema. It is important to note that the validity concept is not as precise as the well-formedness concept. The XML document is either well formed, and then any XML processor is able to read it, or it is not well formed, and then any XML processor is required to reject it. However, an XML document can be valid according to one schema and invalid according to another schema at the same time.

There are many ways to write schemas for XML documents, but only a few are commonly used. XML designers usually select from this triumvirate of schema languages: Document Type Definition (DTD), W3C XML Schema, and RelaxNG. Sometimes they augment their validation with Schematron rules.

Document Type Definition (DTD)

DTD is the oldest XML schema language. Its rules are declared in the same standard as the XML language (Bray et al. 2006b), and many software tools support DTD. It is the only schema language that can be used directly inside XML documents. DTD declarations are used not only for validation purposes, but also as vehicles for providing names to uncommon characters or inclusion of external files.

An important advantage of DTD is its compact and relatively simple syntax, but DTDs also have several disadvantages. Their expressive power is rather limited, and the schema languages discussed below are much more powerful. Another serious problem is caused by the fact that DTDs are not namespace aware. Some methods exist to simulate namespace handling in DTDs, but such solutions are rather complicated and prone to problems. Their syntax is not based on XML notation, so generic XML tools cannot be used for editing DTD documents.

In spite of these shortcomings, at least basic knowledge of DTD syntax is indispensable for everyone seriously working in the XML field. DTD was the only schema language used in the early XML years, and even nowadays it is commonly exploited both for validation and other purposes.

W3C XML Schema

W3C XML Schema is a schema language defined by the W3C (Thompson et al. 2004; Biron and Malhotra 2004). The choice of name for this schema language was very unfortunate. As discussed above, several schema languages can be used for validation of XML documents, and W3C XML Schema is just one of them. Unfortunately, many people do not understand this nuance and often insist on W3C XML Schemas in situations where another schema language would be more suitable.

W3C XML Schema is a powerful validation language that is well supported by software tools, but it has some disadvantages that are important to keep in mind. Its specifications are very complex and difficult to thoroughly comprehend. This unfortunate fact led to many corrections to the original specifications, and many problems are still not fully resolved. Although people are shielded from this complexity by schema-authoring software, they become hostages to their tools. If a tool does not offer some validation rule, it may be very difficult to write it correctly by hand; if two validators return contradictory results, it takes a disproportionate amount of time to decide which one is right. W3C XML Schemas are also rather weak in some commonly encountered circumstances such as the handling of content without strict order found in many textual documents.

It is significant that several flagship W3C activities are not using W3C XML Schemas for their specifications but are using RelaxNG as the schema language of choice (e.g., Scalable Vector Graphics development, http://www.w3.org/TR/SVGMobile12/schema.html).

RelaxNG

In contrast to the previous schema languages, RelaxNG is not a product of the W3C. It was originally developed independently, but later accepted by OASIS, and submitted to a rigorous standardization procedure that resulted in standard ISO/IEC 19757-2:2003 (ISO 2003). From a formal point of view, RelaxNG therefore has higher legal status then the W3C XML Schema, which is just a specification issued by a voluntary industrial group. We are mentioning this fact explicitly because there still exists an unfortunate tendency to reject RelaxNG-based specifications because of the claim that it has spurious legal status.

RelaxNG is a very powerful schema language that is much simpler to use and comprehend then W3C XML Schema. It is particularly suitable for validation of flexible data and for fast prototyping. This makes it the validation language of choice in data mining applications where diverse sources are processed and combined.

Schematron

Schematron is a schema language based on different computing principles than the ones discussed above. It can be combined with any previously mentioned language to provide further validation rules that are difficult or impossible to implement otherwise. Schematron was standardized by an ISO standard (ISO 2006) and is gaining acceptance in many different user scenarios.

Before ending our discussion of schema languages, it is necessary to mention a very important point relevant to data mining. Although neither RelaxNG nor Schematron influence understanding of XML documents, W3C XML Schema and DTDs can cause unexpected surprises. These latter specifications enable attribution of default values to the XML document, so the result of processing an XML document with or without accompanying schema can be diametrically different.

PROCESSING OF XML DOCUMENTS

Any data processing must start with the reading of data into the computer memory. Thanks to rigid XML rules, a lot of programming code can be reused for reading diverse documents. Experience gained with such reuse evolved into standardized procedures for accessing document data.

SAX Processing

When speed is important or the memory requirements are high, SAX processing is usually used. SAX is not officially standardized, as it is a result of community effort (http://www.saxproject.org/), yet it is one of most commonly used XML technologies. In SAX processing, XML documents are continuously read from an external channel (file, database, web connection), and any time the SAX parser encounters element start-tags, end-tags, or textual data, it sends information about the event to the processing application. After sending the data, the parser immediately forgets about them (drops them from computer memory), so there is no limit on the size of documents that can be processed with SAX parsers.

SAX parsers also never return back to already processed data or look ahead for some value needed for current processing. The computation is therefore very fast, and the best parsers can read many megabytes of data in a few seconds, but there is a price to be paid for this speed and absence of memory limits. Programmers are required to create all necessary coding structures by themselves from information sent by SAX parsers, and this often requires a nontrivial programming effort. If ultimate efficiency is not a necessity, it is usually better to let a generic piece of software create more elaborate structures in computer memory—a data model.

DOM (Document Object Model)

XML data can be organized in computer memory in many different ways. XML is used for many purposes, and it is not surprising that there is no optimal organization that fits all circumstances. Yet all these data models share many similar features, so it is possible to create models that are applicable in many different scenarios.

The data model most commonly used is the Document Object Model (DOM) defined by the W3C in a series of standards (see http://www.w3.org/DOM/DOMTR). The W3C DOM model is both platform and language neutral. This means the model is specified in abstract terms and can be translated to particular programming languages such as Java, Python, and JavaScript. The authors of these standards worked very hard to make such translations possible to many different languages of variable expressive power. The specifications therefore do not use possibilities offered by some modern programming languages, and this fact makes processing via W3C DOM quite cumbersome. Nevertheless, the universality of its usage makes up for this restriction. Large amounts of code for the processing of W3C DOM data structures were already written and can be reused in personal projects, and it is quite easy to find programmers experienced in DOM programming.

Applications that require a lot of data in memory or have specific programming requirements may benefit from custom-made data models. Data mining applications, with their large volumes of data, often belong to this class of software, but this does not mean W3C DOM is entirely unusable. Some implementations of the DOM model do not require all data to reside in memory and allow very large documents to be processed.

In real-world XML software applications, a SAX parser is commonly used to feed data into computer memory, where the data is organized into W3C DOM structures. In this standardized way many thousands of hours of programming work are saved thanks to code reuse and the possibility of using familiar techniques. However, there is one particular area of XML processing where an even more general approach can be used, called XSLT transformations.

XSLT Transformations

One of the most common tasks in XML programming is transformation of one data format to another. XML technologies hugely benefit from the fact that a specialized language exists that is particularly suitable for this task: XSLT (eXtensible Stylesheet Language Transformations). In this language, an experienced programmer can write in a few hours quite complex translation software that would typically require days of work in other programming languages. The first version of XSLT recommendations (Clark 1999) has been widely implemented, and XSLT 1.0 processing software can be found in modern Internet browsers. The power of XSLT was substantially increased in version 2.0 and recently approved (Kay 2007). The possibility of data transformation to a format suitable for the application in hand, and the option to change the format many times during development, without significant programming effort, is a very important selling point of XML technologies.

A large part of XSLT processing power lies in its usage of powerful tools for searching and selection of information from XML documents. These tools makes XML a very useful instrument for persistent storage of data.

XML DATABASES

When people are talking about databases they usually mean some complex software systems for storage and fast retrieval of huge amounts of data, but the meaning of term *databases* is much broader. It is a label for any structured collection of data in which big software systems represent just one, albeit very important, subcategory.

We have already discussed the capabilities of XML for handling semistructured data, and when considering the preceding definition it becomes obvious that in reality almost any XML document is in fact a database. However, even if we consider databases in a more restricted sense, as collections of many records sharing a common structure, there are some scenarios where XML beats any competition.

The strength of XML in real-world applications becomes apparent when we leave the world of rigid information structures, where databases are commonly encountered in daily practice (ranging from bank accounts to basic bibliography databases), and we try to capture evolving and slightly fuzzy data, a very common scenario in science and other human endeavors. Anybody with some experience in database programming will attest to the fact that continuous changes in the structure of relational databases and the creation of entry fields for data that occurs infrequently, represent both management and programming nightmares.

Introduction of a new data item into data stored in XML format is much easier. A new element or attribute can be introduced (possibly in a new namespace) without difficulty, provided that the processing code is written with this possibility in mind. Such software will not break after encountering an unknown element, and the processing of information containing this new element can be implemented in an appropriate time.

There is no limit on the size of XML data. Many database vendors have implemented XML support to their products. A new and very powerful language for querying XML databases—XQuery (Boag et al. 2007)—has been adopted and for many tasks, and this language can be more expressive than SQL. XML databases are often implemented on top of established relational ones, so performance is not a problem.

XML MARKUP LANGUAGES

As the preceding text demonstrated, XML technologies offer a very powerful set of tools to simplify data processing, accelerate developments, and improve the quality of final products. However, tools are just one, albeit important, requirement for selection of a particular technology. For successful data mining, it is particularly important to understand how well a document model captures the information presented in the data.

Knowledge of the strengths and weaknesses of individual languages is therefore very important for selection of a suitable strategy. XML documents can appear in all stages of data processing. The original documents may be written in an XML language where programmers have to decide whether it is better to process them directly or if it is more expedient to transform them into some other format. XML is also suitable for capturing intermediate results and for the presentation of results. Hence, the importance of understanding XML data handling cannot be overstressed.

The expressive power of selected languages is of paramount importance. If a language is not capable of capturing all available information, its usage is problematic. This is not a problem with formats used internally to a document, as new elements can be added when need arises, but when providing export formats, it is necessary to live with the fact that some information may be lost. Although achieving high expressiveness is an obvious design aim, it should be always kept in mind that the proposed XML format should be easy to parse. The language selected and its internal formats applied should be fit for purpose and tailored to suit its specifications.

Development of a new language, both for internal and external use, is a very complicated task, and it pays to use functionality from existing languages. Many current XML languages are unfortunately monolithic and often difficult to reuse. The authors commonly implement afresh parts of the language that can be successfully borrowed from specialized domain-specific languages. XML namespaces offer the opportunity for natural inclusion of other languages without these complications. Before attempting to design a new language, everybody should read Tim Bray's article "Don't Invent XML Languages" (Bray 2006) to get some realistic insight. Yet there are many legitimate cases for inventing your own languages, including a very typical case of development for internal prototyping. If a decision is made to start a new development, some useful general rules should be followed.

Many times during design of an XML language it is necessary to decide if elements or attributes should be used for given information. The choice may often reflect personal taste without any deleterious effects, but sometimes the wrong choice can be detrimental to further development. Attributes are suitable for capturing simple information without internal structure. XML documents with attributes are usually easier to read by human readers, and SAX processing code for attribute parsing is simpler than for elements. However, the elements are easier to extend, and they can be used to store multiple instances of particular data. Because of their greater flexibility, elements should be preferred in case of doubt.

Versioning

Both data and requirements are evolving, and when writing data mining applications or designing a new markup language, this is necessary to bear it in mind. It is impossible to get everything right at the first attempt; experience brings new needs and the discovery of unexpected problems. Therefore, during development, the meanings of existing elements and attributes can change, or new ones may be introduced while others are simultaneously depreciated. The hierarchical relations can evolve as elements gain new children or move to other parts of an XML document.

It is therefore important to think early on about a versioning system to keep track of changes and indicate what kind of syntactic constructs can be expected. Version changes can be indicated either by namespace changes or by usage of an attribute or an element dedicated to the version specification. Usage of namespaces for version indication is not common, and there is a good reason for this. As discussed above, the element name consists of two parts: namespace and local name. If some namespace in a document is changed, then all elements from this namespace change their names. Software written for previous versions will not recognize any of these elements, even if a majority of them remain unchanged from the previous version. If, for example, the version change simply signals the introduction of a single seldom used attribute because of the insignificant change, the software become useless.

The naive solution of adding this attribute without indicating a change in version is fraught with problems. Data processing often includes validation steps, and obviously if a document contains this new attribute, the whole document will be rejected as invalid. Switching the validation off may be dangerous. The programmer could make some assumptions because of efficiency considerations (although usually without real need). If some programming code relied on a fixed maximum number of attributes, addition of a new attribute might derail the computing logic with unpredictable consequences.

Because of these considerations, version numbers are usually provided by an attribute on the root element, and namespaces are kept unchanged between versions (or they change only if the language changes significantly).

Flexibility

Some authors of XML languages feel that flexibility contributes to the strength of the language, but flexibility is not the same thing as expressiveness. If the same information can be encoded by several means, it usually adversely affects readability and causes a headache for programmers who must take care of all possible variants. If the syntactic variants are not easily convertible to some common basis, then a combinatorial disaster occurs, as almost any document written in the language may contain a combination of features that were not previously tested.

Flexible documents also restrict the usefulness of the tools used for editing of XML documents. If the order of elements is fixed, editors can make suggestions about expected input and provide appropriate entry fields. In flexible formats a list of variants may be displayed, but as these lists are usually long, their usefulness is restricted.

STANDARDS

Standardization is obviously of paramount importance to data mining. Programming resources are scare, and data locked in unsupported formats can be lost forever. However, whereas the basic idea that standards are useful is generally shared, the real-world situation is far from ideal. In a perfect world, standards would be developed first and agreed upon by all interested parties. With these standards in hand, interoperable products would be developed, but experience shows most standards created this way utterly fail. Standardization is a complex process, and experience is only gained through stepwise development and usage.

In reality several ideas compete for success, and companies that invested a lot of money in development of their own variant are reluctant to throw away that investment even if competing ideas are technically superior. The success of technology therefore depends on many reasons other than technical merit. Some understanding of the standardization process helps with estimation of potential risks and separation of technical merits from marketing hype.

Most standards used by the XML community are issued by the W3C and by OASIS, but many of them started as private activities of small groups without any official ties (SAX, Relax, Schematron). Specialized markup languages are usually backed by professional organizations or by inventors of the language.

WORLD WIDE WEB CONSORTIUM

The W3C was founded in October 1994 by Tim Berners-Lee at the Massachusetts Institute of Technology, Laboratory for Computer Science in collaboration with CERN, but it was later transformed to a different and very powerful body. The consortium issues many important standards that are available at http://www.w3.org/TR/. The standardization process (Jacobs 2005) ends with voting, and because the voting rights are derived from financial contributions, big companies dominate the voting. Fortunately, protection from patent litigation is provided because any recommendation accepted by the W3C can be implemented on a royalty-free basis (Weitzner 2004).

ORGANIZATION FOR THE ADVANCEMENT OF STRUCTURED INFORMATION STANDARDS (OASIS)

OASIS, a not-for-profit consortium, plays an important role in XML standardization. Although it does not have the same media visibility as the W3C, its processes are more democratic than the W3C and less dependent on the financial strength of participants (http://www.oasis-open.org/who/policies_procedures.php). Because the OASIS standardization processes can produce ISO standards, their final products are of a higher legal status than the W3C products. Both RelaxNG and Schematron are examples of such OASIS processes.

XML AND CHEMICAL DATA MINING

XML is a young technology, and its basic infrastructure is still evolving, as witnessed by the number of recommendations published each year. Scientific XML publishing needs this infrastructure and at the same time requires specialist knowledge. Thus, it will take several years before scientific markup languages reach maturity and even longer before they become ubiquitous. Reliable prognosis of further developments concerning survival and future changes of languages is very difficult, so we will concentrate on general observations and provide links to relevant literature and Internet resources for readers interested in more details.

CHEMICAL STRUCTURES AND REACTIONS

Formats for the capture of molecular structures and reactions are obviously of paramount importance in chemical data mining. Fortunately, a lot of experience had been amassed before the invention of XML, so it was not necessary to start from the beginning. Computational chemists are used to using connectivity tables, and XML provides a very convenient syntax for their formal specification.

One of the major setbacks of previously used text formats is their reliance on white-space characters as text separators and problems with extensibility if new demands call for format expansion. In properly designed XML languages, white-space characters loose any formatting role, so accidental introduction or deletion of white-space characters during editing or programming is no longer problematic. Moreover, the hierarchical structure of XML documents provides a convenient mechanism for annotating individual atoms or bonds with different properties without significant complications. In XML introduction of new attributes or elements (possibly in different namespaces) will not break properly programmed software applications. Many problems will be discovered by validation against a variety of schemas, and valid documents can be further processed with powerful software libraries.

Chemical reactions are straightforwardly described by the introduction of several elements and attributes. Because the hierarchical nature of XML enables the natural grouping of molecules, the separation of reactants and products is simplified. The stoichiometry, yields, or reaction conditions can be conveniently expressed with additional attributes or elements.

This simplicity, unfortunately, vanishes when more vague concepts are introduced. For example, the capture of all relevant information is an elusive aim when structures are not precisely defined, when dealing with electron delocalization, or when complex relations exist between reactants and products. Although current chemical formats provide mechanisms to address these problems, a general satisfactory solution has not been yet found.

Graphical presentation of reactions and compounds represents another complex area. A markup language may attempt to provide some mechanisms for providing presentational hints, but as long as the presentational format is just an add-on of a chemical language, its expression power will be weak. Much broader possibilities are offered by synergistic use of a format directly targeted for graphics. Scalable Vector Graphic (SVG), a W3C XML format (Ferraiolo et al. 2003) for two-dimensional graphics, is gaining popularity and supported by an increasing number of applications including Internet browsers. The power of expressing graphical data with SVG is demonstrated by an open-source chemistry editor BKChem (http://bkchem.zirael. org/) that combines SVG with chemistry data in its native storage format.

CHEMICAL MARKUP LANGUAGE (CML)

Chemical Markup Language (CML) (Murray-Rust and Rzepa 1999, 2003) has played a prominent role in the early developments of chemistry markup languages. Its development started several years before the final XML recommendation was officially issued (Murray-Rust et al. 1995), so CML was very important for XML's evolution. CML creators and their collaborators have further expanded the original specification. CMLReact (Holliday et al. 2006) extends the applicability of CML into chemical reactions, and CMLSpect (Kuhn et al. 2007) extends it into analytical techniques. Nevertheless, CML has several weaknesses that prevent it from becoming a really powerful chemical syntax. Because the language is very flexible, and provides several means for capturing chemical data, including methods external to the specification, it is difficult to write fully compliant software. Although generation of a CML-compliant file is not difficult, it is an impossible task to properly read all possible CML files.

The CML convention attribute is especially pernicious. With this attribute, it can be specified that a given notation should be understood according to rules given in some external specification. Although some people consider this behavior to be an advantage, as it enables simple transformation from any format to CML, in reality this dramatically reduces the usefulness of the language. There are no predefined conventions in CML, and there is no formal way to describe these conventions or to specify their locations. The seriousness of the problem can be demonstrated on this potential scenario: Today you save your compounds in CML format with a function offered by the structure editor. In 10 years' time, you will try to mine your data, but the editor used 10 years earlier used its own conventions for bonding between atoms. These conventions are now hard to find or lost forever. Even if they are found, you have to write specialized software to identify and interpret the notation. The same problem may occur with legacy chemistry databases or outputs of computational software.

CML has been tailored for use with the Java programming language. Some features of CML could be more elegant and easier to implement if a more abstract approach was chosen. CML also offers the possibility of encoding information in a non-XML way. Whereas experienced programmers have no difficulties writing code to parse white-space-separated fields or to handle x2 and x3 attributes as the same information (CML has different notations for the x and y axes in two-dimensional and three-dimensional contexts), such work is tedious and increases the chance of programming errors. Moreover, it severely limits the possibilities of performing data validation with standard XML tools.

CML is the best known of the XML notations for capture of structural data, but several other formats use XML-based syntax. The Protein Data Bank (PDB) (http:// www.wwpdb.org/) is the single worldwide repository for macromolecular structure data. A representation of the Brookhaven PDB is available in an XML format called PDBML (Westbrook et al. 2005). PDBML provides a way to export structures and information about them from a relational database. Another database that offers download of its data in a native XML format is PubChem (http://pubchem.ncbi.nlm.nih.gov/), which provides information on the properties and biological activities of small molecules, covering several million compounds.

PHYSICAL MEASUREMENTS

Nowadays any chemistry endeavor produces a massive amount of physical data. Measurements range from routine operations performed daily in synthetic laboratories to very sophisticated measurements with very expensive apparatuses. Only a very small part of the data is effectively used. Most individual data points are never published, and the published data are often provided in forms not suitable for automatic processing.

Standardization of formats for capturing physical data would dramatically accelerate progress in the natural sciences, not only because of new discoveries stemming from mass processing of available data, but also because of the time saved by highly qualified people, which could be more fruitfully invested elsewhere.

The creation of a useful language for physical data capture is a demanding endeavor. The format must deal with immense variability of apparatuses and measurements, so expertise in many different fields is required. It is not easy to separate particularities of given measurements from more general features. It is therefore not surprising that the progress has been rather slow, but current developments in areas such as ThermoML and AnIML are very promising (as described below).

Special attention is required when handling units of measure. A standardized methodology for units would enable the writing of software libraries usable in many various scenarios and thus thoroughly tested for consistency and software errors. Because a single error in one unit conversion can lead to great financial losses and safety concerns, this is an area that deserves the utmost attention. The recent development of UnitsML by the National Institute of Standards and Technology (NIST) promises to bring much-needed unification.

Handling of uncertainties is another important area that requires careful thinking. Measurements without uncertainty considerations lose most of their utility, and if a format does not provide a mechanism for their lossless capture, its usage is problematic. Some languages have an element or attribute for a basic uncertainty specification, but ThermoML is the only language that deals with uncertainty seriously.

ThermoML

ThermoML (Frenkel et al. 2006) is a language for the storage and exchange of experimental, predicted, and critically evaluated thermophysical and thermochemical property data. The language aims to cover essentially all relevant properties, and it deals thoroughly with uncertainties (Chirico et al. 2003). The standard, which has been developed as an IUPAC project (2002-055-3-024) and become an IUPAC recommendation (Frenkel et al 2006), provides detailed controlled vocabulary for the field. Several high-profile journals already provide data in this format (http://trc.nist.gov/ThermoML.html), and if this trend continues in the future, introduction of ThermoML may be rightfully considered as one of the milestone achievements in the rich history of thermodynamics research.

Although ThermoML represents excellent work, it has several XML-related problems. For example, there are some differences between published schema and IUPAC recommendations (http://www.iupac.org/namespaces/ThermoML/update_061218. html). In these cases the published schema provides the authoritative answer. It is also unfortunate that ThermoML documents do not use XML namespaces. The absence of thermodynamic namespaces makes recognition of ThermoML code needlessly complicated. Furthermore, ThermoML language could also benefit by focusing on thermodynamics data and capturing associated data such as bibliographical citations in some established language.

AnIML (Analytical Information Markup Language)

The target area of AnIML (http://animl.sourceforge.net/) is the handling of analytical data. The teams participating in AnIML development, sanctioned by ASTM under subcommittee E13.15, have not published a final recommendation yet, but the language promises to become one of the most important XML languages in the chemistry field.

In the heart of the language lies a flexible core applicable to many analytical techniques, as described by the W3C XML Schema. Specific needs of individual techniques are addressed in "Analytical Technique Definition Documents" that are XML files specifying particular parameters. The attempt to separate core features from particularities in a modular way is laudable. Other specification authors would benefit from using this approach.

UnitsML (Units Markup Language)

NIST has played a leading role in the development of both ThermoML and AnIML specifications. It is not surprising that people from the same organization stand behind the development of a common language for the expression of units—UnitsML. The language is still in development (http://unitsml.nist.gov/), and recently an OASIS technical committee has been created that will steer final stages of standard development. AnIML development will probably include UnitsML in the near future (Jopp et al. 2006).

Several languages are focusing on more specialized areas of physical measurements. MatML (http://www.matml.org/) has been developed for the exchange of materials information. The HUPO proteomics standards initiative (http://www.psidev.info/) is developing several standards including mzML (http://www.psidev.info/index.php?q=node/257), a standard for encoding raw data from mass spectrometers that builds on the previous formats DataML (http://www.psidev.info/index.php?q=node/80) and mzML (Pedrioli et al. 2004). Another HUPO standard, PSI MI XML, provides syntax for the description of molecular interactions (http://www.psidev.info/index.php?q=node/31).

MATHEMATICAL EXPRESSIONS

Many areas of chemistry need some notation for the capture of mathematical expressions and complex symbols. Representation of mathematics is very complex, and we strongly advise against individual development of these languages. Fortunately, there exists a mathematical XML language: MathML, by the W3C (Carlisle et al.

2003). MathML has mature specifications that meet most requirements for chemistry. It thoroughly addresses the capturing of both mathematical meaning and its graphical representation, including a sophisticated means for the handling of symbols. Moreover, MathML is supported by several software packages including several Internet browsers.

SBML (SYSTEMS BIOLOGY MARKUP LANGUAGE)

SBML (Hucka et al. 2003) is a machine-readable format for describing qualitative and quantitative models of biochemical networks that use MathML for the specification of mathematical formulas. It defines a subset of the language to be used in the SBML, using a widely applicable standard, without the need to implement all features of the MathML specification. The excellent documentation provided for SBML provides an example worth following (http://sbml.org/documents/).

The SBML homepage (http://sbml.org) provides an impressive list of software implementing the notation. The site offers downloads of software libraries that enable easy incorporation to several programming languages (C, C++, Java, Python) as well as mathematics packages (Matlab, Mathematica), and this accessibility has significantly contributed to its success.

RESOURCE DESCRIPTION FRAMEWORK (RDF)

Introduction of XML formats was a very important step toward better intercomputer communication, but it is not a miraculous solution to all problems. Not every possible relation is easily expressed in XML (Wang et al. 2005), so specifications usually contain many implicit assumptions that are not properly formalized. The Resource Description Framework (RDF) provides a very powerful yet simple model for this formalization (Manola and Miller 2004). In this framework, any information is transformed to basic units called triplets that are combined to map the available information. This unifying mechanism can be used to express hierarchical vocabularies for domain knowledge description, as in RDF Schema (Brickley and Guha 2004) or its extension, Web Ontology Language (OWL) (Smith et al. 2004), both standardized by the W3C.

BioPAX (http://biopax.org/), the group developing a common exchange format for biological pathways data, uses OWL. BioPAX can capture molecular binding interactions and manage small molecules (represented by InChI). Another example of RDF usage in chemistry is provided by the CombeChem project (Taylor et al. 2006).

Abstract models are very powerful, but at the same time, the abstractness is their Achilles' heel. Many people have problems operating on such a high level of abstraction and have difficulties grasping the concept. A middle ground where XML-based formats provide the scaffold for expressing the basics and RDF or other mechanisms are used to specify relations may provide more suitable practical applications.

CONCLUSIONS AND PERSPECTIVES

XML has started the second decade of its official existence, and already in its early years it has deeply influenced many areas of information processing. Its promise for scientific information exchange is only slowly being fulfilled, but recent developments are very positive. XML formats are replacing their text predecessors, so an understanding of XML technologies is a must for anyone interested in data mining applications.

Scientific XML formats are still rapidly evolving, and it will take some time before stability is reached, but this does not mean they are unusable today. XML already provides several advantages over its predecessors including the availability of tools that simplify conversion to newer versions of XML when necessary.

Science is diverse, and this is reflected in the fragmentation of scientific XML formats that are independently developed for each field of science. Hence, the same concepts are reinvented several times, with slightly different semantics, making it difficult to recycle software code across fields. The explosion of the number of elements and attributes introduced in scientific XML languages is staggering, making it is very difficult to gain a deeper understanding of relevant specifications. This unfortunate fact partially stems from the complexity of the problem. Much could be gained if development were more modularized in a concerted manner. XML offers mechanisms for seamless integration of small domain-specific formats to languages of high expressive power.

Research in chemical formats also suffers from a dearth of open communication channels. Unlike the bioinformatics field, where there is a prevalent practice of openness including open access to most literature articles, the chemistry field is relatively insular. Limited availability of information experts restricted the development of the chemical format field, where most contributors are much stronger in chemistry than in software architecture and programming.

Nowadays, *Semantic Web* is a much-hyped expression that is used (and misused) by many people from different areas of human endeavor. On Semantic Web computers we will be able to autonomously search for different information items and process them and provide answers to questions that currently require many hours of effort by human experts. However, before we can connect the individual pieces of information, we have to find and transform them to computer-readable form. Unfortunately, this is not a trivial task.

Progress achieved in natural language processing is amazing, but any average human being so far surpasses any natural language processing software capabilities. In a specialist field, such as chemistry, that combines quantitative and qualitative textual and visual information with a lot of intuition, the appearance of software that would understand chemistry without external help is very improbable in the foreseeable future. Partial understanding of information captured in some formalized way is much easier, and current computing hardware and software is capable of supporting evaluation of such information on a mass scale.

XML is not the only existing technology for such formalization, but it is a very suitable candidate. If data are captured in XML syntax, then it is much easier to find software tools and people capable of using these tools. An XML-based format should be a primary consideration when choosing a way to annotate information. Unless some serious deficiencies are discovered, XML is currently the tool of choice. Choosing an XML-based approach does not automatically mean its syntax adequately captures the area of scientific interest. Although XML simplifies programming tasks, and offers a myriad of existing tools, it does not substitute for careful thinking. The success of a language depends primarily on its ability to capture data from its domain and requires that a carefully designed language is not mutilated by inferior XML design.

REFERENCES

Biron, P. V., and Malhotra, A. 2004. XML Schema Part 2: Datatypes Second Edition. Available at http://www.w3.org/TR/xmlschema-2/.

Boag, S., Chamberlin, D., Fernández, M. F., Florescu, D., Robie, J., and Siméon, J. 2007. XQuery 1.0: An XML Query Language. Available at http://www.w3.org/TR/xquery/.

Bray, T. 2006. Don't Invent XML Languages. Available at http://www.tbray.org/ongoing/When/200x/2006/01/08/No-New-XML-Languages.

Bray, T., Hollander, D., Layman, A., and Tobin, R. 2006a. Namespaces in XML 1.0 (Second Edition). Available at http://www.w3.org/TR/xml-names.

Bray, T., Paoli, J., and Sperberg-McQueen, C. M. 1998. Extensible Markup Language (XML) 1.0 (First Edition). Available at http://www.w3.org/TR/1998/REC-xml-19980210.

Bray, T., Paoli, J., Sperberg-McQueen, C. M., Maler, E., and Yergeau, F. 2006b. Extensible Markup Language (XML) 1.0 (Fourth Edition). Available at http://www.w3.org/TR/2006/REC-xml-20060816.

Brickley, D., and Guha, R.V. 2004. RDF Vocabulary Description Language 1.0: RDF Schema. Available at http://www.w3.org/TR/rdf-schema/.

Carlisle, D., Ion, P., Miner, R., and Poppelier, N. 2003. Mathematical Markup Language (MathML) Version 2.0 (Second Edition). Available at http://www.w3.org/TR/MathML2/.

Chirico, R. D., Frenkel, M., and Diky, V. V. 2003. ThermoML: an XML-based approach for storage and exchange of experimental and critically evaluated thermophysical and thermochemical property data. 2. Uncertainties. *Journal of Chemical and Engineering Data* 48:1344–1359, DOI: 10.1021/je034088i.

Clark, J. 1999. XSL Transformations (XSLT) Version 1.0. Available at http://www.w3.org/TR/xslt.

Ferraiolo, J., Fujisawa, J., and Jackson, D. 2003. Scalable Vector Graphics (SVG) 1.1 Specification. Available at http://www.w3.org/TR/SVG11/.

Frenkel, M., Chiroco, R. D., Diky, V., Dong, Q., Marsh, K. N., Dymond, J. H., Wakeham, W. A., Stein, S. E., Königsberger, E., and Goodwin, A. R. H. 2006. XML-based IUPAC standard for experimental, predicted, and critically evaluated thermodynamic property data storage and capture (ThermoML) (IUPAC recommendations 2006). *Pure and Applied Chemistry* 78:541–612, DOI: 10.1351/pac200678030541.

Holliday, G. L., Murray-Rust, P., and Rzepa, H. S. 2006. Chemical markup, XML, and the World Wide Web. 6. CMLReact, an XML vocabulary for chemical reactions. *Journal of Chemical Information and Computer Sciences* 46:145–157, DOI: 10.1021/ci0502698.

Hucka, M., Finney, A., Sauro, H. M., Bolouri, H., Doyle, J. C., Kitano, H., Arkin, A. P., Bornstein, B. J., Bray, D., Cornish-Bowden, A., Cuellar, A. A., Dronov, S., Gilles, E. D., Ginkel, M., Gor, V., Goryanin, I. I., Hedley, W. J., Hodgman, T. C., Hofmeyr, J. H., Hunter, P. J., Juty, N. S., Kasberger, J. L., Kremling, A., Kummer, U., Le Novère, N., Loew, L. M., Lucio, D., Mendes, P., Minch, E., Mjolsness, E. D., Nakayama, Y., Nelson, M. R., Nielsen, P. F., Sakurada, T., Schaff, J. C., Shapiro, B. E., Shimizu, T. S., Spence, H. D., Stelling, J., Takahashi, K., Tomita, M., Wagner, J., and Wang, J. 2003. The systems biology markup language (SBML): a medium for representation and exchange of biochemical network models. *Bioinformatics* 19:524–531, DOI:10.1093/bioinformatics/btg015.

ISO. 2003. ISO/IEC 19757-2:2003. Information technology. Document Schema Definition Language (DSDL). Part 2. Regular-grammar-based validation. RELAX NG. Available at http://standards.iso.org/ittf/PubliclyAvailableStandards/index.html.

ISO. 2006. ISO/IEC 19757-3:2006 Information technology. Document Schema Definition Language (DSDL). Part 3. Rule-based validation. Schematron. Available at http://standards.iso.org/ittf/PubliclyAvailableStandards/index.html.

Jacobs, I. 2005. World Wide Web Consortium Process Document. Available at http://www.w3.org/Consortium/Process/.

Jopp, R., Roth, A., Linstrom, P. J., and Kramer, G. W. 2006. Incorporating Units Markup Language (UnitsML) into AnIML (Analytical Information Markup Language). *Abstracts of Papers of the American Chemical Society* 231:57-CINF.

Kay, M. 2007. XSL Transformations (XSLT) Version 2.0. Available at http://www.w3.org/TR/xslt20/.

Kuhn, S., Helmus, T., Lancashire, R. J., Murray-Rust, P., Rzepa, H. S., Steinbeck, C., and Willighagen, E. L. 2007. Chemical Markup, XML, and the World Wide Web. 7. CMLSpect, an XML vocabulary for spectral data. *Journal of Chemical Information and Computer Sciences* 47:2015–2034, DOI: 10.1021/ci600531a.

Manola, F., and Miller, E. 2004. RDF Primer. Available at http://www.w3.org/TR/rdf-primer/.

Murray-Rust, P., and Rzepa, H. S. 1999. Chemical markup, XML, and the Worldwide Web. 1. Basic principles. *Journal of Chemical Information and Computer Sciences* 39:928–942, DOI: 10.1021/ci990052b.

Murray-Rust, P., and Rzepa, H. S. 2003. Chemical Markup, XML, and the World Wide Web. 4. CML Schema. *Journal of Chemical Information and Computer Sciences* 43:757–772, DOI: 10.1021/ci0256541.

Murray-Rust, P., Rzepa, H. S., and Leach, C. 1995. Chemical Markup Language. *Abstracts of Papers of the American Chemical Society* 210:40-Comp Part 1. Available at http://www.ch.ic.ac.uk/rzepa/cml/.

Pedrioli, P. G.A., Eng, J. K., Hubley, R., Vogelzang, M., Deutsch, E. W., Raught, B., Pratt, B., Nilsson, E., Angeletti, R. H., Apweiler, R., Cheung, K., Costello, C. E., Hermjakob, H., Huang, S., Julian R. K., Jr., Kapp, E., McComb, M. E., Oliver, S. G., Omenn, G., Paton, N. W., Simpson, R., Smith, R., Taylor, C. F., Zhu, W., and Aebersold, R. 2004. A common open representation of mass spectrometry data and its application to proteomics research. *Nature Biotechnology* 22:1459–1466.

Smith, M. K., Welty, C., and McGuinness, D. L. 2004. OWL Web Ontology Language Guide. Available at http://www.w3.org/TR/owl-guide/.

Taylor, K. R., Essex, J. W., Frey, J. G., Mills, H. R., Hughes, G., and Zaluska, E. J. 2006. The Semantic Grid and chemistry: experiences with CombeChem. *Journal of Web Semantics* 4:84–101.

Thompson, H. S., Beech, D., Maloney, M., and Mendelsohn, N. 2004. XML Schema Part 1: Structures Second Edition. Available at http://www.w3.org/TR/xmlschema-1/.

Wang, X., Gorlitsky, R., and Almeida, J. S. 2005. From XML to RDF: how semantic web technologies will change the design of 'omic' standards. *Nature Biotechnology* 23:1099–1103, DOI: 10.1038/nbt1139.

Weitzner, D. J. 2004. W3C Patent Policy. Available at http://www.w3.org/Consortium/Patent-Policy/.

Westbrook, J., Ito, N., Nakamura, H., Henrick, K., and Berman, H. M. 2005. PDBML: the representation of archival macromolecular structure data in XML. *Bioinformatics* 21:988–992, DOI: 10.1093/bioinformatics/bti082.

Part III

Trends in Chemical Information Mining

7 Linking Chemical and Biological Information with Natural Language Processing

Corinna Kolářik and Martin Hofmann-Apitius

CONTENTS

INTRODUCTION

One of the great challenges in natural language processing (NLP) for life sciences is the identification and the extraction of relationships between chemical entities and biomedical entities with the goal of establishing links between chemical and biological information. The ability to do so would allow for systematic screening of the literature for biological activities of chemical compounds and thus is one of the core aims of text mining activities in both the academic world and the pharmaceutical industry. However, biology and chemistry are still quite distinct worlds that communicate their results in very different ways. Biologists and medical researchers, for example, tend to describe chemical compounds by using brand names. In the world of chemistry we prefer the far more informative, unambiguous International Chemical Identifier (InChI) descriptor or other suitable nomenclature-based designators for naming chemical entities.

However, a chemistry-centric view does not naturally favor a system-oriented interpretation of the biological (side) effects of a chemical compound. Therefore, dealing with the task of linking biological with chemical information also means dealing with quite different name spaces, scientific viewpoints, and communities. Automated analysis of the scientific literature for relationships that link biological or medical entities with chemical entities is an interdisciplinary scientific and technological challenge.

An automated system that analyzes the scientific literature for links between biological and chemical information should thus be able to support researchers in satisfying two fundamental information needs:

- Finding relevant documents that contain information about chemical compounds and biological concepts of interest (information retrieval task)
- Finding specific information about chemicals that are related to biological concepts by an explicit relationship from many documents (information extraction task)

Whereas in the first task the focus is on the comprehensive identification of biological and chemical entities in text, the second task deals with the identification and extraction of explicit relationships between chemical and biological entities. These two tasks are relevant to both academic an industrial research.

At a more technical level, the task of linking biological and chemical information can be divided into at least four high-level subtasks:

- Named entity recognition (NER)
  - Recognition of biomedical entities in text (genes, proteins, allelic variants, clinical phenotype descriptions, etc.)
  - Recognition of chemical entities in text (drug names, chemical descriptors, registry numbers, common and brand names, etc.)
- Information retrieval: From statistical analysis of full text indices to semantic search based on entity classes defined by NER
- Information extraction: Analysis of part-of-speech and identification of expressions that link biological and medical entities with chemical entities
- Presentation of the retrieved results for supporting navigation and mining in the scientific literature

The perspectives of such technology for the life sciences are quite promising, but as we will show in this chapter, the technology developed so far for linking chemical and biological information is still in its infancy.

RECOGNITION OF BIOMEDICAL ENTITIES IN TEXT

SHORT METHODOLOGICAL OVERVIEW

Biological entities in scientific text comprise all named entities that represent genes (nucleic acid sequences), mutations (deletions and insertions), and allele variants of genes (e.g., single nucleotide polymorphisms [SNPs]), transcripts (mRNA), chromosomes, proteins, and protein complexes. In the medical literature, named entities encompass anatomical terms, disease names and disease classifications, clinical descriptors of diagnostic procedures, and clinical treatment descriptions including drug common and brand names. In the following, we will briefly discuss the heterogeneous tasks faced by an automated system for the recognition of biomedical entities in scientific text.

In principle, four different basic approaches for named entity recognition can be distinguished:

- Systems based on dictionaries
- Systems based on (expert) rules
- Systems based on machine learning
- Systems combining two or all three of these approaches

Systems based on dictionaries rely on the availability of terminologies in a given domain. In biology, databases are a good source for gene and protein name terminologies. The advantage of dictionary-based systems is that it is comparably simple to create lists containing synonyms and term variants. Rule-based approaches make use of rules generated by human experts that are aware of the specific syntax and name space used in a domain. A simple example for a rule-based approach is the identification of enzymes based on the suffix "-ase" (e.g., peptidase) or the identification of diseases based on the suffix "-itis" (e.g., gastritis). Machine learning–based systems are dependent on the availability of relevant text corpora annotated by experts. During a training phase, the computer program is taught how a certain class of entities "looks like" and "learns" the probabilistic rules required to identify other members of that class of entities.[1] Both rule-based and machine learning–based systems have the intrinsic ability to discover new entities that were previously unknown and therefore are not present in dictionaries.

GENE AND PROTEIN NAME RECOGNITION

The name space of proteins and genes in higher eukaryotes is rather diverse. Some known human genes and proteins have more than 100 synonyms in the literature.[2,3] The situation is made even more complicated by the invention and use of acronyms and spelling and permutation variants by biologists and medical researchers. Moreover, ambiguous use of acronyms and gene names that sometimes resemble common words adds to the already existing complexity of the named entity recognition task.[4] Systems that identify those types of entities need to deal with all those challenges. Gene and protein name recognition is one of the most advanced text mining applications in the life sciences. The growing number of publications in this scientific area supports this observation.[5] The predominant role of gene and protein name recognition in biomedical text mining is also illustrated by the fact that gene and protein name recognition were central tasks in the two BioCreative critical assessments of text mining in molecular biology[6] and in the JNLPBA workshop.[7]

An inevitable step to the correct identification of protein and gene mentions is the mapping of the identified proteins and genes to unique database identifiers—a process called normalization.

Machine learning approaches for the recognition of gene and protein names have been improved significantly over the past three years, but the identification of synonyms, the resolution of acronyms, and the handling of spelling variants remain significant challenges to this kind of approach. Moreover, gene and protein names identified with the help of machine learning approaches cannot easily map entities in text to entries in databases. Some groups have tried to overcome this by combining dictionary and machine learning approaches.[8]

Systems based on dictionaries and algorithms to deal with permutations, spelling variants, and acronyms[9] are well suited for normalization of named entities that is the mapping of entities in text to entries in databases. Through normalization of textual entities to database entries, the number of attributes associated with a named entity in text can be significantly increased; a gene name has a nucleic acid sequence in European Molecular Biology Laboratory (EMBL);[10] a protein has a structure in Protein Data Bank (PDB).[11] Of course, the ability to link back from database entries to Medline abstracts through the Gene2PubMed list available from National Center for Biotechnology Information (NCBI)[12] may help establish this sort of relationship from the database side. In fact, Gene2PubMed has been used to generate gold standards for corpora containing relevant information on genes and proteins.[13]

The characterization as a target protein might become possible through the combination of textual information with entries from databases such as UniProt,[14] PubChem,[15] or DrugBank.[16] At ChEBI,[17] the European Bioinformatics Institute's database on bioactive small molecules, this integration of chemical information with biological entities is realized through expert curation of entries and introduction to referential links to databases such as UniProt and PDB.

Likewise, an automated approach to link chemical and biological information based on text mining will profit largely from being able to normalize textual entities and to map them to nonredundant referential entries in databases.

Recognition of Information on Mutations

Gene mutations are tightly linked to biological phenomena such as the risk of developing a certain disease or the responsiveness to pharmaceutical treatment. The entire vision of personalized medicine in the pharmaceutical context is closely linked to the genetic makeup of the patient, and the relationship between genetic mutations and pharmacological responses is the subject of the new field of pharmacogenomics. As a substantial portion of the biomedical literature is composed of information on mutations and their association with phenotypes, the identification of information on allele variations (e.g., single nucleotide polymorphisms [SNPs]) from text provides another challenge to information extraction technology.

Although public databases such as dbSNP[18] do contain a large number of known SNPs including their gene and positional information, a significant fraction of SNP mentions in scientific text cannot be mapped to dbSNP entries.[19]

Mentions of mutations can be automatically identified using a combination of gene dictionaries and rule-based approaches for the identification of such expressions.[20] Alternatively, a combination of dictionary-based gene and protein name recognition methods and machine learning–based identification of expressions indicating information on mutations has been developed.[19] Both approaches enable the identification of SNPs that are normalizable but also find SNPs that are not normalizable. At present, it remains unclear to what extent these "non-normalizable" SNPs result from incorrect numbering of nucleotides, lack of submission to dbSNP, or simply referencing to outdated sequence information.

Rule-based approaches are predicated on the analysis of expressions that describe mutations in text; the rule set developed is based on patterns that have been analyzed by human experts. Learning of patterns is the basis for machine learning approaches for SNP detection. Similar to the human expert, the machine program makes use of features (attributes) that are used with different frequency and in different combinations when information on mutations is communicated in natural language.[21]

CONCEPT-BASED IDENTIFICATION OF FUNCTIONAL BIOLOGICAL ENTITIES

In the past years hierarchies such as MeSH[22] introduced for article and book indexing of the National Library of Medicine, the Gene Ontology (GO)[23] and other bio-ontologies for the characterization and classification of genes and gene products (e.g., Panther,[24] TAIR,[25] MapMan ontology for plants,[26] and the Sequence Ontology Project[27]) have been developed. Concepts taken from these hierarchies and ontologies have been used in text mining approaches to map functions to entities.

GoPubMed,[28] for example, provides tagging of Gene Ontology (GO) categories in Medline and allows for mapping of expressions to GO categories.[29] Other ontology-based text mining tools such as Textpresso[30] developed their own categorization system for the analysis of biological knowledge in text. The concept-based identification of biological entities has the advantage that generalizations are possible and similarities can be established at an abstract level without being bound to the instance of an entity.

The concept-based identification of biomedical entities is now being offered by most commercial vendors of text mining solutions and by an increasing number of academic tools.[31] Biomedical concept search is mainly based on publicly available thesauri such as MeSH or GO (see below).

Nomenclatures, terminologies, and controlled vocabularies established in molecular biology, for example, GO, are a basis for NER, but when mining literature, we always have to keep in mind that the use of terms describing the functions of proteins took place before GO was established, and the meaning of certain functional descriptions might have changed over time.[32]

RECOGNITION OF MEDICAL TERMINOLOGY

Medical terminology and thesauri can be detected in a similar fashion to that discussed in the context of biological entity recognition. Approximate search algorithms have been proven to be quite useful for dealing with the variability of expressions indicating the same concept. However, a couple of principle problems exist when trying to identify medical phenotype descriptions and using Medline abstracts as a data source:

- Clinical phenotypes are predominantly described at a categorical level. Clinical phenotypes as we can find them in patient records are rarely found in scientific publications.[33]
- Despite the availability of different coding systems, authors typically do not indicate the referential information on a concept they use. This means references to medical ontologies such as the Foundational Model of Anatomy[34] or a thesaurus such as the International Classification of Diseases[35] are not made explicit in the scientific literature.

Thus, a system for the recognition of medical entities in text has to offer functionalities beyond simple string matching—namely, context-dependent disambiguation—to support the mapping of entities in text to concepts in medical ontologies.

RECOGNITION OF CHEMICAL ENTITIES IN TEXT

Chemical information is generally more proprietary than biological information. Whereas molecular biology and genome research profited immensely from the open access to sequence information and annotations, the world of chemistry—in particular, the chemistry of bioactive molecules—is much more restricted. Although chemistry as a discipline is much older than modern molecular biology, a significant amount of knowledge on bioactive compounds is kept confidential.

Only recently, initiatives have been started to create freely available data sources such as PubChem,[36] ChEBI,[37] DrugBank,[38] and HMDB,[39] to mention some of them. These public databases collect publicly available information on compounds, their structures, their physical formulations (e.g., PubChem Substance), their targets, and their effects on biological processes. However, these databases are far from covering the entire spectrum of chemical information that can be linked to biology and pharmacology.

Therefore, automated methods for the detection and extraction of chemical information in scientific text are required. Although chemical named entity recognition is dealt with in detail in Chapter 3 of this book, we briefly summarize the principles underlying chemical named entity recognition. We also briefly elucidate on the need for reconstruction of chemical information from chemical structure depictions, as detailed in Chapter 4.

When discussing chemical named entity recognition, distinguishing between the different nomenclatures types used for chemical entities in text is inevitable. Basically, there are five classes:

- Common names used in communication between chemists
- Common names used in marketing (brand names)

- Systematic names or nomenclature names (e.g., International Union of Pure and Applied Chemistry [IUPAC])
- Chemical structure representation formats (e.g., InChI, Simplified Molecular Line Entry System [SMILES])
- Catalog and registry numbers

Similar to the identification of information on gene mutations in text, we can differentiate between chemical names that can be listed and enumerated in a dictionary (e.g., common names and brand names) and those cases where the potential name space is almost infinite and thus cannot be enumerated (e.g., IUPAC expressions and terms with an IUPAC-like structure).

For brand names and other common names, dictionaries composed of reference names, synonyms, and brand names have been of great help for the detection of named chemical entities in scientific text. However, even though normalization might be possible for a significant number of "important" compounds, these compounds are typically the ones that are known best, and therefore the link between chemical and biological information can be established readily. These compounds can also be found regularly in the public chemical databases such as PubChem or ChEBI.

A significant fraction, however, of the documents in the scientific literature dealing with chemical entities and their biological effects are not composed of trivial names for the compounds under investigation. For the automated analysis of the chemical named entities in these publications, we need to use other methods. In principle, it should be possible to use rule-based approaches to identify IUPAC names (and other forms of IUPAC-like expressions), in particular, because the IUPAC name construction itself is based on rules. However, IUPAC names are neither unambiguous, nor can they easily be checked automatically for compliance with IUPAC nomenclature rules. In fact, most IUPAC-like expressions in patent literature seem to be not compliant with the IUPAC nomenclature, and cannot easily be converted into structures.[40]

IUPAC-like expressions, true IUPAC nomenclature names, and InChI and SMILES representations of chemical compounds are well suited for detection by machine learning approaches. Conditional random fields (CRFs)[41] and support vector machines have been used for the detection of IUPAC expressions in scientific literature.[42] Other approaches are based on rules sets[43,44] or combinations of machine learning with rule-based approaches.[45] All these approaches have in common that they face one significant problem: the "name-to-structure" problem.

The three-dimensional structure is the most unique description of a chemical compound. That is why chemical entities should be compared on the basis of their structure as represented in a connection table, not on their common or nomenclature name. Comparison of structures, however, requires that mentions of chemical entities in text are translated into connection tables; this is typically done by name-to-structure (N2S) tools. On a conference on chemical information in Sitges (International Chemistry Information Conference [ICIC]) 2007), preliminary data on attempts at benchmarking N2S tools were reported.[46] Although this analysis is preliminary and care should be taken to avoid drawing conclusions that are not supported by the analysis, these data suggest that the N2S tools currently available are correctly converting only between 30% and 50% of all named entities.

CHEMICAL ENTITY RECOGNITION IN CHEMICAL STRUCTURE DEPICTIONS

Research results about properties, reactions, and syntheses of chemical compounds, especially novel findings, are mainly communicated through structure depictions placed into text (see Chapter 4). Because the machine readability of chemical structure information is lost during printing or electronic publishing without additional meta-information behind structure depictions, the chemical information is only readable for human experts. This problem poses a significant challenge for automated mining of chemical information in scientific literature. Scientists from different organizations have been tackling the problem of reconstructing chemical structure information from depictions. Four approaches to chemical structure reconstruction have been reported at the time of writing this review:

- Kékulé[47]
- CliDE[48]
- ChemoCR[49]
- OSRA[50]

Because Chapter 4 of this book deals with the challenges of chemical structure reconstruction from images, we refer to this chapter for details on the technological and algorithmic basis for software dealing with this challenge.

Currently, there are no published reports on attempts at combining text information with information extracted from chemical structure depictions. However, the CliDE tool is sold in a professional version together with a text mining module,[51] which indicates that the problem of data aggregation from textual and image data is being recognized. Preliminary work done in our group has demonstrated that chemical structure information reconstructed from chemical structure depictions can be used for the annotation of patents.[52]

FUNDAMENTALS OF NLP

Natural language processing (NLP) studies the problems of the automated understanding of natural languages and is a subfield of artificial intelligence and computational linguistics.[53] NLP systems basically analyze the structure (syntax) and semantics of text. In the process text is split into single units (tokens) as a first step and then a certain grammatical category (i.e., noun, adjective, verb, etc.) is assigned to each, which is called part-of-speech tagging. To discover the meaning of text or to extract information of interest, it is necessary to identify the fundamental information carriers—nouns or noun phrases—which can be single words or multiword terms. For this process, called chunking, patterns are used that describe the general composition of such phrases. Domain-specific terms such as proteins, genes, or chemicals that are nouns often have a different morpho-syntactic structure compared to common nouns. The recognition of named entities by methods introduced in the previous sections supports the correct identification of those nouns. At the same time, we have a mapping of terms to an entity class, such as protein or chemical compound. For example, the MetaMap[54] software developed by NLM discovers noun phrases in text and maps them to concepts used in the UMLS Metathesaurus.[55]

Finding relationships between chemical and biomedical entities by NLP methods requires the identification of general grammar patterns (e.g., <nounphrase1> verb <nounphrase2>) being the basic syntactic structure used for the description of relations. Incorporating semantics provided by entity classes of interest allows for the definition of lexico-syntactic patterns (e.g., <drug> verb <protein>) for the relationship extraction.[56] The manual creation of such patterns is very labor intensive, so automated techniques for the learning of the pattern structure are used as support. Machine learning methods combined with statistical analysis of the syntax structure are used to extract general rules for a given set of manual annotated examples.[57,58] One of the challenges is the recognition and mapping of different variants describing the same type of relation, for example "**azothymidin** *is* an **inhibitor of reverse transcriptase**" and "**azothymidin** *inhibits* **reverse transcriptase**." Here, the inhibition relationship between the entities is encoded either by a verb or a noun.

Another point is the position or the context in which a named entity or relationship occurs in text. It reflects its importance for the information transported by the whole document. The analysis of the document structure itself (e.g., dividing it into title, abstract, introduction, results, etc.) helps weight and structure the found information.

IDENTIFICATION OF RELATIONSHIPS THAT LINK BIOMEDICAL ENTITIES WITH CHEMICAL ENTITIES

Methods that support the identification of relations between entities can be divided into those that use statistical analyses of the occurrence of named entities in whole documents, paragraphs, or sentences and those that rely on a deeper analysis of the underlying syntactic and semantic structure at the sentence level of every document. Co-occurrence is the simplest type of relation between two named entities identified by the previously described methods. It has the advantage that no effort has to be spent on analyzing its underlying syntactic structure. This is one of the reasons why co-occurrences have been widely used in academic and commercial text mining tools. Surprisingly, co-occurrence as a basic type of relationship already suffices to identify and to represent a significant portion of the existing relationships in scientific text.[59] However, the method is not dedicated solely to getting the information about how the entities are related to each other (e.g., "effects of aspirin overdose include renal failure, pyrexia, ... ," "aspirin inhibits cyclo-oxygenase1 and cyclo-oxygenase2").

NLP techniques provide the basis to extract this kind of more detailed information. In the past years approaches have been developed that are mainly focused on the biomedical domain to extract protein–protein interactions[60] and gene–disease relations.[61] They are based upon the correct recognition of the named entities taking part in the relationship. Dedicated patterns have to be developed to identify all the phrasal constructs that are indicative of relationships between chemical and biomedical entities available in text being of interest for academic researchers and the pharmaceutical industry.

Various types of relationships between biomedical and chemical entities can be distinguished by the grade of information they provide. We give some examples of the most important and—in our opinion—most informative classes of relationships that occur in biomedical text in the sense of their meaning. They can be divided into nonspecific relationships that describe loose associations between entities and those carrying very specific information.

The association type of relationships ("has something to do with") contains inherently vague information; it can only provide statistical evidence or phenomenological evidence, but typically this relationship does not provide a mechanistic explanation. We find this type of associative relationship frequently in publications dealing with genetic epidemiology analyses ("**Association between** the CYP17, CYPIB1, COMT and SHBG polymorphisms and serum sex hormones in post-menopausal breast cancer survivors") or reports on molecular and sometimes complex effects of drugs ("PKA-induced resistance to tamoxifen **is associated with** an altered orientation of ERalpha towards co-activator SRC-1").

Relationships of higher specificity (and most likely also higher expressivity) are often used to describe molecular interactions that are understood to a much higher level of detail. A class of relationships that can readily be identified is the class of enzymatic reactions. Typical expressions used for this type of relationship are "glucuronidation_of" or "proteolytic_cleavage_of." We found that relationships specifying enzymatic reactions are mainly used in the biological domain; in some cases we find them in publications that belong to the pharmacological domain.

Another class of expressions with pharmaceutical relevance that can be identified by a rather simple pattern is the type of relationships that display the following structure "<drug / compound name>-induced..." or "<drug / compound name>-mediated... ." This type of expression indicates a relationship between a chemical compound or drug and a biological or medical entity and thus represents a typical statement of pharmaceutical interest. However, biological entities must not be mixed up with drug and compound names. In many cases, we find expressions such as "estrogen-receptor-mediated transcription ..." or "PKA-induced resistance ...," and in both cases the acting entity is not a chemical entity (a drug or compound) but a biological entity (a protein).

Yet another class of relationships with relevance for pharmaceutical research is the induction, repression, or any other form of regulation of biological processes by chemical compounds such as natural ligands or drugs. Efforts need to be made to automatically extract such information and provide them well organized by visualization tools as support for database curators and researchers.

STATE OF ACADEMIC RESEARCH IN APPLYING NLP TECHNIQUES TO LINK BIOLOGICAL AND CHEMICAL ENTITIES

One of the first works that dealt with extracting information on chemical entities from text and relating this to another entity (in this case chemical structure similarity) was published by Singh et al. in 2003.[61a] In contrast to the steep increase of publications dealing with biological entities, the growth of reports on chemical entity recognition systems is lagging behind in the literature. This finding is *not* paralleled by the development of commercial solutions, indicating that once again the proprietary nature of pharmaceutical chemistry information is delaying development in this sector. Our own group has adopted the ProMiner system, originally developed for biological entity recognition in text, to deal with chemical entities. This adapted version of ProMiner has been used to search chemical trivial names in Medline abstracts and other sources of scientific text. Through co-occurrence analysis, we were able to identify relationships between proteins and drugs. However, in the area of pharmaceutical research, co-occurrence as a relationship seems not to be sufficiently informative, as pharmaceutically relevant relationships tend to be rather complex; just think of *allosteric inhibition* or *cooperativity* as concepts describing relevant pharmaceutical relationships at the molecular level.

More recently, in the course of the EU-integrated project ANEURIST,[62] we have developed a data-mining environment that supports mining in various literature sources (Medline abstracts, full text scientific papers, and patents). The entity types that can be detected in text span from genes and proteins via mutations and chromosomal locations to medical terminology, ontology terms, and drugs names, as well as nomenclature names for chemical compounds (IUPAC-like expressions). The system maps the identified entities of the classes protein, gene, chemical compound, and disease to respective databases or thesaurus or ontology entries (normalization). The system has been used for the identification of genes associated with a disease (intracranial aneurysm) and benchmarked against a recently published expert review on genes involved in this disease.[63] Figure 7.1 shows the interface in its version for the ANEURIST project with a search for SNPs of genes associated with breast cancer, which are mapped on their representation in the dbSNP database. Figure 7.2 shows the document view with the various tagged entity classes.

The EBIMED tool,[64] developed at the European Bioinformatics Institute, allows for simultaneous analysis of biological, medical, and chemical entities and uses the drug dictionary of MedlinePlus.[65] However, as with our early experiments with ProMiner, EBIMED is mainly based on co-occurrence as a rather basic relationship.

Regarding chemical named entity recognition, new momentum in the development of technology to detect named chemical entities in text comes from the work of the groups participating in the OSCAR project. This team has allied with the UK Royal Chemical Society and is using the system for the analysis of chemical information in textual sources. However, so far, no evidence can be found from the OSCAR website or literature searches that this system is being extended to deal with biological or medical entities.[66]

In the area of pharmacogenetics and pharmacogenomics, two groups have reported on the successful application of text mining technology for linking biological and chemical information. Chang and Altman have demonstrated that a combination of a co-occurrence approach and a machine learning approach for the classification of relationships is able to collect and classify information on drug–gene (and protein) relationships with satisfactory results.[67] Thomas Rindflesch's group at the National Library of Medicine has recently published on the extraction of semantic predications or relations from Medline citations for pharmacogenomics.[68] For the classification of relationships, the group modified the ontological representation of pharmacogenomics concepts in UMLS.

In an impressive body of work, researchers from IBM published on using the UIMA framework for the assembly of a complex workflow composed of services for biological, medical, and chemical entity recognition and relationship extraction. This publication is discussed in more detail in the following section.

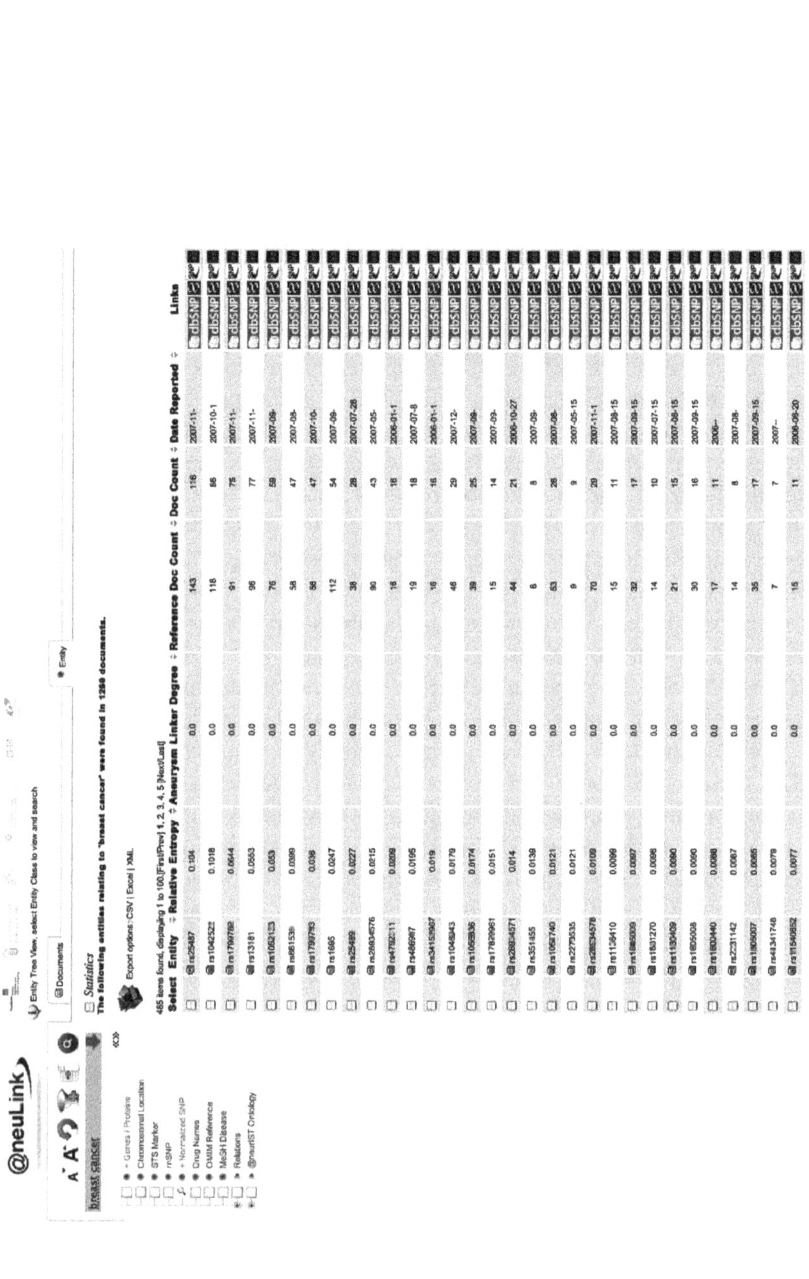

FIGURE 7.1   Selection of information automatically extracted from Medline on polymorphisms (SNPs) of genes that are significantly associated with breast cancer in the scientific literature.

FIGURE 7.2 *A color version of this figure follows page 146.* Document view with tagged color-coded entities representing chromosomal locations, drug names, protein and gene names, STS markers, OMIM references, entities from the dedicated @neuIST disease ontology, SNP information (normalizable and nonnormalizable), MeSH disease terms, and simple relation terms. A gene of interest (BRCA1) has been highlighted.

Our own group has recently published an approach to extract pharmacological property classes that are directly related to chemical compounds from Medline abstracts. The system is based on a commercial solution, the TEMIS platform for text analysis, using so-called Hearst patterns. Patterns indicative of information on the biological activity of compounds were identified and used to analyze Medline abstracts and the unstructured text field in DrugBank for drug classification statements. We identified additional information not yet contained in DrugBank annotation fields.[69]

A very interesting application of text mining technology for hypothesis generation in the area of toxicogenomics has been reported recently.[70] In this work, text analysis of Medline abstracts was used to generate keyword "fingerprints," which were then used to analyze patterns in microarray gene expression data. Dependent on treatment of cells in microarray experiments, keyword profiles of publications dealing with the substance or the substance class mentioned in text were correlated with gene expression data. The patterns detected gave raise to new hypotheses on the toxicogenomics regulation events caused by these compound classes.

UNSTRUCTURED INFORMATION MANAGEMENT ARCHITECTURE (UIMA)

In 2005, IBM released a framework for unstructured information management to the open source community: the Unstructured Information Management Architecture (UIMA) framework.[71] UIMA is meant to be "an open, industrial-strength, scaleable and extensible platform for creating, integrating and deploying unstructured information management solutions from combinations of semantic analysis and search components." At the technical level, the UIMA framework provides a standardized format for the handling of annotations of unstructured information sources (including multimedia data) and a JAVA framework to that is distributed as a software development kit (SDK). According to IBM, the goal of UIMA is "to provide a common foundation for industry and academia to collaborate and accelerate the world-wide development of technologies critical for discovering the vital knowledge present in the fastest growing sources of information today." The underlying idea is that it is rather unlikely, that the requirements of application domains for text and multimedia analytics can be comprehensively addressed by monolithic solutions. An open, service-oriented architecture like UIMA is meant to provide a standardized framework that allows a broad community of industrial and academic developers to assemble complex workflows for text analytics by deploying individual tools as services in the UIMA framework.

Some academic initiatives in the area of text mining and natural language processing have adopted the UIMA concept recently,[72] and some commercial vendors of text mining solutions have done so, too.[73] Of course, IBM itself is a major user of UIMA.[74]

IBM has delivered one of the early examples for usage of the UIMA framework to build a solution for biomedical NLP that is able to link biological and chemical information. The IBM BioTeKS (Biological Text Knowledge Services) system[75] is composed of a large set of annotators for the analysis of biomedical text (Medline, patents, full text journals). BioTeKS provides solutions for the syntactic analysis of the documents, identifies a variety of named entities belonging to previously described chemical and biomedical entity classes, and can extract relations between those.

- Tokenization of the text (identification of the smallest units forming an expression; includes in this case lemmatization)
- Part-of-speech tagging
- Shallow parsing for the analysis of syntactic units (defines part of speech as a noun or verb)
- Dictionary lookups for lexical information (basically identifying word stems that match an authority, in this case MeSH terms)
- Dedicated annotation of UMLS concepts in text (dictionary lookup specialized for UMLS; comprises rule-based disambiguation of terms)
- Recognition of chemical nomenclature names (rule-based approach for the recognition of chemical fragment strings)
- Recognition of drug names and associated dosage qualifiers (combination of dictionary and rule-based approach for the identification of drug names and the extraction of dosages information)
- Mapping of terms to ontology concepts (adding information about the hierarchical position of a concept to the matching term in text)
- Mapping of lexical terms and complex semantic categories to expressions in text (this is a term categorizer based on machine learning)
- Extraction of relations (based on shallow parsing, this annotator identifies syntactic clauses containing noun and verb phrases)

BioTeKS thus is composed of almost all functionalities to analyze biomedical literature including patents for mentions in text of chemical and biological as well as medical entities and the extraction of their relationships. The advantage of the implementation of this system as a UIMA-based, service-oriented architecture is that annotators can be added or replaced depending on the needs of the user of the system. In essence, the only service that is missing from the complex BioTeKS is an analysis service for the annotation of chemical structure depictions.

NAVIGATION TOOLS FOR NLP LINKING OF BIOLOGICAL AND CHEMICAL INFORMATION

A very important aspect of mining in unstructured data sources is the presentation of the results to the user, typically an expert in a given domain. A steadily growing number of tools for the presentation and navigation of results coming from information extraction approaches is being made available; in this review we will focus on two of them: Cytoscape[76] and AliBaba.[77]

Cytoscape is a bioinformatics software platform for visualizing molecular interaction networks and integrating these interactions with gene expression profiles and other state data. Cytoscape has been designed in a modular way, and one of the core features of the system is the extension of functionality through plugins. After about five years of work by the open source consortium driving this project, plugins are available for network and molecular profiling analyses, new layouts, additional file format support, and connection with databases. Anyone can develop plugins using the Cytoscape open Java software architecture, and the development of new plugins by the community is strongly encouraged. Cytoscape has become the standard environment for data analysis in systems biology projects; several commercial text mining solutions use Cytoscape as the visualization front end (see section on commercial solutions below).

AliBaba is a visualization and querying front end for the analysis of text documents. Ali Baba parses PubMed abstracts for biological objects and their relations. The tool visualizes the resulting network in graphical form, thus presenting a quick overview over all information contained in the abstracts. For many relations, Ali Baba searches for simple co-occurrences in the same sentence. For protein–protein interactions and cellular locations of proteins, a more sophisticated strategy is used in addition.

COMMERCIAL SOLUTIONS FOR NLP-BASED LINKING OF BIOLOGICAL AND CHEMICAL INFORMATION

Over the past five years a growing number of commercial NLP tools for the biomedical and chemical domain have reached the market. Some of these solutions are offered by companies such as SAS, IBM, and SPSS, using well-established statistical modeling and mining techniques. Other solutions have been developed by smaller software companies founded between 2000 and 2004 (e.g., BioAlma, Linguamatics, BrainWave, and TEMIS). In the following, we will briefly introduce some commercial solutions, which support mining strategies linking biological and chemical information. As a note of caution, the product presentations below have been supplied by the vendors. The following is therefore by no means a critical review of the technology behind these tools.

According to what is discussed by users of commercial systems, open issues are the ability to handle huge numbers of documents and the processing of large amounts of full-text documents. Connected to these issues is the question of the performance of these solutions and their scalability. Complete analysis of documents by NLP systems is a rather computer-intensive task, as the systems need to analyze the structure of each sentence in a document, and, in particular, patent literature can be very demanding at the syntax level. Finally, we do not yet see any system appearing at the horizon that would be able to combine text analysis with image analysis, something that is of high relevance for approaches that aim at linking biological and chemical information.

The following list introduces commercial vendors of text mining solutions in alphabetical order. The authors do not claim that the selection of vendors is comprehensive; the selection rather represents the willingness of vendors to provide information upon request.

> **BioAlma** (http://www.bioalma.com): Bioalma's text mining tool is called AKS (Alma Knowledge Server). It identifies and extracts a wide range of concepts, including biological and chemical entity types, from the scientific literature.

The system uses dictionaries, statistical methods, and very specific heuristics for the identification and extraction of each concept type; the combination of these approaches enables the system to distinguish between different entities. Information on genes and chemical substances is normalized to database entries wherever possible. Figure 7.3 shows how the BioAlma AKS identifies drugs mentioned in the context of a target protein (COX2).

BrainWave (http://www.brainwave.in): BrainWave's tool, Text Miner (see Figure 7.4), is composed of modules for both biology and chemistry. Text Miner can work with a variety of document formats (TXT, PDF, and HTML formats) as input. On the NLP side, Text Miner is composed of modules for named ntity recognition, parts of speech tagging, rule-based entity relationship extraction, and a network builder. A specific gene–drug relationship extractor module is part of Text Miner; the identification of biological and chemical entities and their relationships is based on a combination of dictionary-based approaches, rule-based-approaches, and pattern-recognition methodologies. The system supports mapping of chemical entities in text to two-dimensional structures in public data sources.

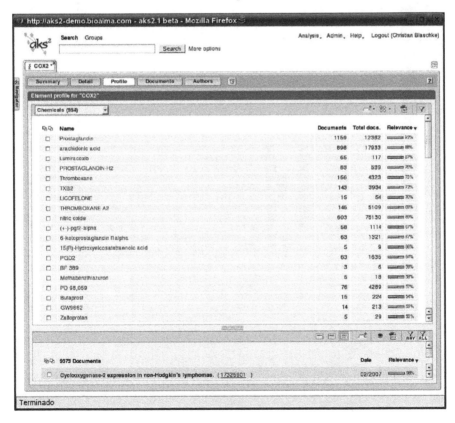

FIGURE 7.3 A section of results obtained for the gene Cox2. Shown are the chemical compounds related to the gene. The relevance gives an indication of the strength of the relationship as it is found in the literature.

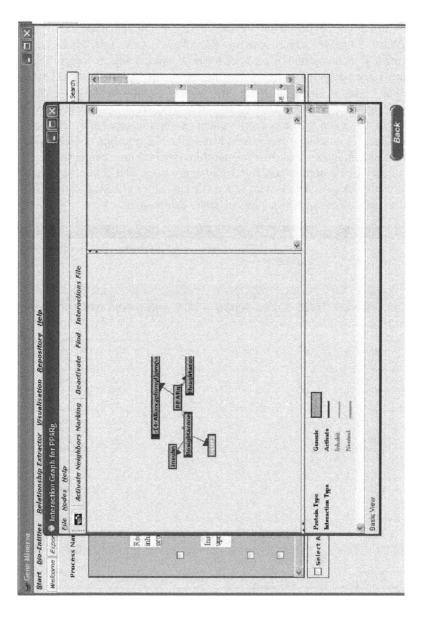

FIGURE 7.4 Interaction graph generated by Text Miner displaying relationships between the PPARg protein, binding drugs, and insulin as a factor influencing the pathway which with the drug interferes.

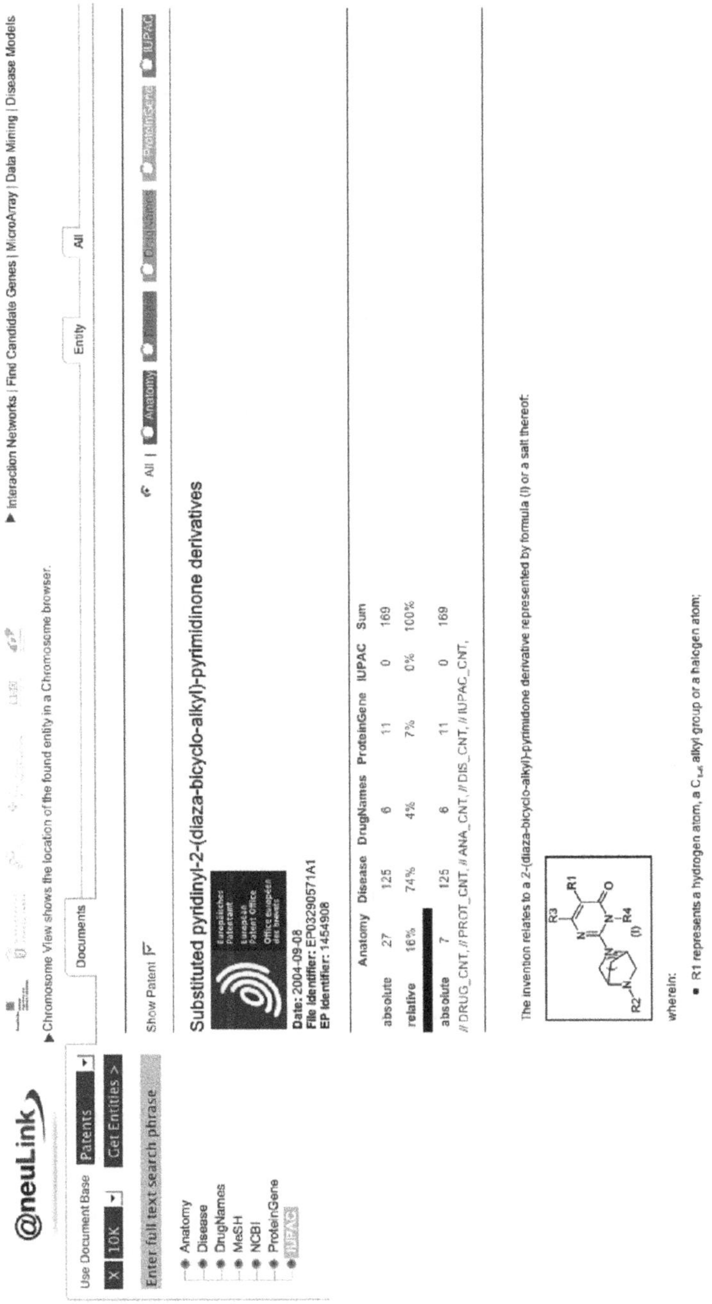

FIGURE 7.5 Overview of entity types tagged in a full-text patent document. Absolute numbers of concepts in the patent as well as the relative percentage of a given concept class versus the total number of concepts used in the patent are provided. Tagged entity classes are anatomical concepts, disease terms (specialized disease ontology), drug names (common names), MeSH terms, NCBI taxonomy terms, protein and gene names, and IUPAC-like expressions.

Fraunhofer SCAI (http://www.scai.fraunhofer.de): Fraunhofer SCAI has developed a suite of tools for biomedical and chemical named entity recognition based on dictionaries (ProMiner) and machine learning (IUPAC and SNP recognition tools). Recently, a tool for the reconstruction of chemical information from chemical structure depictions (ChemoCR) has been added. All these tools are enabled to work in distributed systems (clusters); their output is stored in a persistence layer with a web-based front end (SCAIview; see Figure 7.5). Named entities that can be recognized are gene and protein names, mutations (normalizable and nonnormalizable SNPs), chromosomal locations, genetic markers, MeSH terms, drug names, OMIM reference terms, any sort of controlled vocabulary (e.g., GO), and defined types of relationships of disease risk factors. SCAIview supports complex querying of Medline abstracts as well as full-text documents (e.g., patents). An integration of chemical structure reconstruction through ChemoCR for the analysis of patents is possible. The system accepts documents in ASCII, XML, HTML, TXT, and PDF format; ChemoCR accepts images in TIFF, JPG, and any bitmap format.

Linguamatics (http://www.linguamatics.com/): Linguamatics I2E (Interactive Information Extraction) uses NLP-based querying to extract relevant facts, relationships, and quantitative data from large document collections. Semantic search capabilities are enhanced by plugging in domain knowledge in the form of taxonomies, thesauri, and ontologies. Query results are presented in a range of structured forms, including tables with highlighted hits and direct links to source documents. Linguamatics states that I2E is capable of searching millions of documents, for example, querying over the entire Medline corpus, and handles ontologies of millions of terms. I2E provides a technology platform, which is applicable to many domains and can reveal insights across different types of information. Figure 7.6 illustrates the ability of I2E to extract quantitative values associated with a drug (in this case IC50 values linked to Taxol).

SAS (www.sas.com): The text mining tool offered by SAS is called Text Miner (see Figure 7.7); the tool is part of the larger SAS Enterprise Miner system. Following the general competence profile of SAS, the tool is based mainly on statistical approaches in text analysis; dedicated domain knowledge (e.g., proprietary pharma ontologies, gene and protein dictionaries, annotated corpora for training of machine learning tools) is not part of the solution but can be used by the system if available. Application areas for the text miner in the life science domain are composed of the "identification of authors and co-authors for specific therapy" and approaches to optimize clinical studies.

TEMIS (http://www.temis.com/): Luxid (see Figure 7.8), the text mining solution offered by TEMIS, is a service-oriented, modular architecture built on the UIMA[78] framework. The identification in, and extraction from, text of biological and chemical information is mediated by dedicated modules for biological and chemical entities. Relationships between biological and chemical entities are identified through dedicated rule-based and

▼Paclitaxel	1 microM	▼2|6149300	1	Taxol inhibits acetylcholine-stimulated catecholamine secretion (IC50: approximately 1 microM) and 45Ca++ uptake.	2
		8104621	1	Veraparnil (10 microM), and taxol (IC50 = 1 microM) prevented VBL uptake and evoked VBL diffusion from vesicles when added after VBL uptake had reached steady state.	2
	0.003-0.150 microM	110626810	1	Paclitaxel was the most potent agent (IC50 = 0.003-0.150 microM); sulindac sulfide, NDGA, and 13-cis-retinoic acid had intermediate potency (IC50 = 4-80 microM), and cisplatin and exisulind were the least potent (IC50 = 150-500 microM).	2
	0.03-0.38 microM	111036469	1	The extent of response markedly varied between the different cell lines, although chromophilic RCCs exhibited a more pronounced response to Taxol (IC50: 0.03-0.38 microM) than clear cell RCCs (IC50: 0.01-36.69 microM).	2
	0.05, 0.06, 0.03 and 0.04 microM	8521379	1	In human OVCAR-5, PANC-1, H-125 and rat 3924A cells, for taxol the anti-proliferative IC50 was 0.05, 0.06, 0.03 and 0.04 microM, respectively; for tiazofurin IC50 = 8.3, 2.3, 1.8 and 6.9 microM.	2
	0.08 microM	8581880	1	In the bone slice assay, taxol (0.1-0.001 microM) dose-dependently inhibited bone resorption with an IC50 of 0.08 microM.	2

Taxol inhibits acetylcholine-stimulated catecholamine secretion (IC50: approximately 1 microM) and 45Ca++ uptake. The inhibitory effects of both taxol and vinblastine on secretion are rapid in onset (approximately 1 min) and reversible. Taxol and vinblastine (5 microM) exert little or no inhibitory effect on catecholamine secretion induced by 1) the nonreceptor mediated secretagogues K+, Ba++ or veratridine or by 2) the receptor-mediated secretagogues histamine or bradykinin. Similarly, taxol and vinblastine had no effect on K+-induced 45Ca++ uptake into chromaffin cells. The inhibitory effects of taxol and vinblastine during a secretory challenge are specific for cholinergic receptor-mediated 45Ca++ uptake and catecholamine release and prevent receptor-mediated membrane depolarization. These results do not support a role for microtubules either in the exocytosis event or in granule transport during an initial secretory challenge. The results would be consistent with either an interaction of microtubule protein with the acetylcholine receptor or a direct action of the drugs on the acetylcholine receptor.

McKay D B
Schneider A S
In Vitro
Journal Article

FIGURE 7.6 I2E-mediated extraction of specific numerical data (here, IC50 values) from scientific literature. Shown is an example result for paclitaxel (the active ingredient of taxol), where the most frequently mentioned concentration is 1 µM, with evidence from two documents.

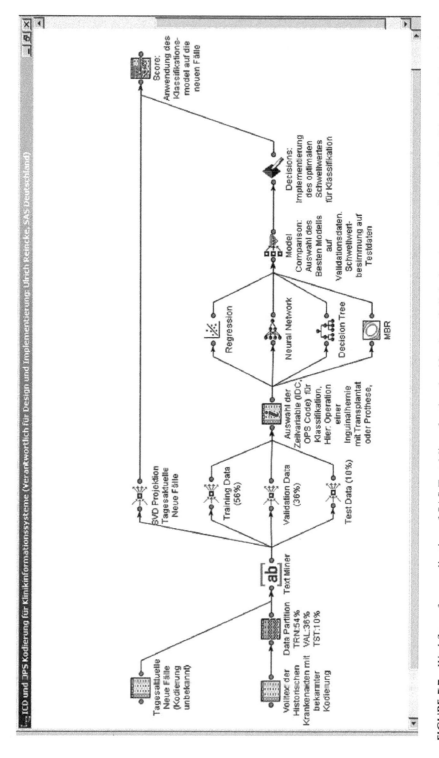

FIGURE 7.7 Workflow of an application of SAS Text Miner for the classification of clinical patient records containing Diagnosis Related Group (DRG) codes.

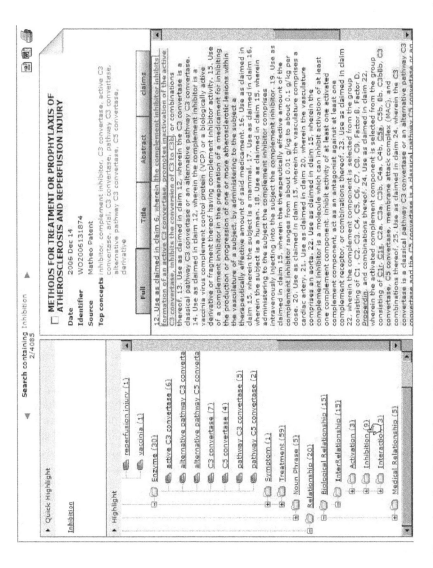

FIGURE 7.8 Luxid user interface showing the results of a search in a patent document. Various entity types and relationships are being categorized and made navigable. An extract of the top concepts used in the document is provided with the document information.

pattern-based methodologies; Luxid provides an environment for the rapid generation of domain-specific grammar. The system supports the inclusion of competitive intelligence information in mining strategies that link chemical and biological information. The system normalizes to database entries wherever possible.

SUMMARY AND CONCLUSION

Text mining in the life science literature currently envisages a growing interest from both the pharmaceutical and biotechnology industries, as well as from the academic research community. However, advancements in technology development seem to be confined to application domains. Named entity recognition in biology has reached a quality that now allows using text mining systems for database content production, whereas in the domain of chemistry we do see significant challenges in the area of named entity recognition yet to be addressed.

The application of NLP technologies to link between the domain of biology and the domain of chemistry has just been started in the academic research communities. First reports on analyzing co-occurrences of biological and chemical entities in text have been published during the last two years. So far, most of the analyses published have been done on Medline abstracts. It remains to be shown whether and to what extent the analysis of full text documents enables linking between chemical and biological information at a higher degree of granularity. At this point, it is noteworthy to recall that one category of chemical information, namely the chemical information residing in structure depictions, is only available from full-text documents. The combination of text mining with the automated analysis of chemical structure depictions has not been demonstrated yet. We envision that this combined extraction of information will be nontrivial.

In the area of commercial information extraction systems, we observe a strong "pull effect" from the pharmaceutical and biotechnology industry. Automated methods of information extraction are regarded as tools for increasing productivity of human specialists in this industry. This increase of productivity can be achieved through improved retrieval of relevant information, combination of competitive and scientific intelligence, and extraction, aggregation, and presentation of factual statements and relationships from heterogeneous unstructured text sources. In particular, the vision of automated mining of patents at a substantial degree of granularity is appealing to the industry. The ability to generate flexible, customizable views of biological, chemical, and medical content extracted from the scientific literature (including patent literature) is one of the major aims of text mining activities in the pharmaceutical industry. In the ideal case, profiles of biological entities and networks, medical indication areas, and chemical structure spaces could be defined as individualized search patterns. Alerting services would inform the user about new information available on these entities and their relationships.

So far, commercial text mining systems that link biological and chemical information through NLP seem to be ahead of academic developments. However, we have only limited information on the performance of these commercial systems. In particular, the rather sophisticated NLP approaches are computer-intensive, and currently

FIGURE 7.2 Document view with tagged color-coded entities representing chromosomal locations, drug names, protein and gene names, STS markers, OMIM references, entities from the dedicated @neurIST disease ontology, SNP information (normalizable and nonnormalizable), MeSH disease terms, and simple relation terms. A gene of interest (BRCA1) has been highlighted.

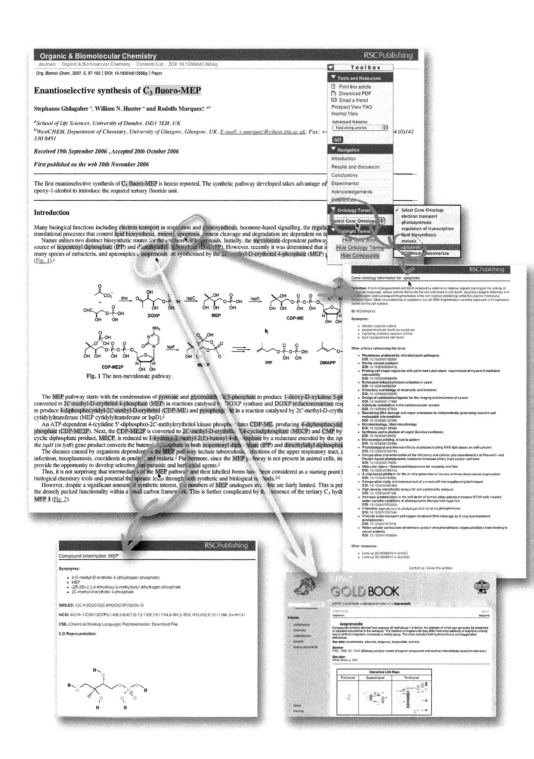

FIGURE 8.1 Semantic mark-up of content linking to additional data sources.

no reports are available on true large-scale analysis of relationships between chemical and biological entities in patents or other full-text corpora. We believe that at present the systems linking biological and chemical information using NLP have already demonstrated their potential, but they are not yet productive in the sense of the vision sketched in the beginning of this chapter.

The UIMA initiative launched by IBM has the potential to facilitate the construction of complex document analysis workflows. Services from academic laboratories as well as commercial tools could be integrated in one workflow as long as all tools can work with the standardized common analysis structure. A growing number of academic research groups and commercial development teams are adopting UIMA; we therefore expect to have a growing number of UIMA annotator services available in the public domain and in commercial solutions built on UIMA.

NLP approaches combining biological and chemical information have just been started during the past years; we should therefore feel encouraged to assume that the majority of work in this area still lies ahead of us.

NOTES AND REFERENCES

1. Cohen, K.B., Hunter, L. 2008. Getting started in text mining. *PLoS Computational Biology.* 4(1):e20.
2. Fundel K, Zimmer R. 2006. Gene and protein nomenclature in public databases. *BMC Bioinformatics.* 7:372.
3. Liu, H., Hu, Z.Z., Torii, M., Wu, C., Friedman, C. Quantitative assessment of dictionary-based protein named entity tagging. *Journal of the American Medical Informatics Association.* 13(5):497–507.
4. Spasic, I., Ananiadou, S., McNaught, J., Kumar, A. 2005. Text mining and ontologies in biomedicine: making sense of raw text. *Briefings in Bioinformatics.* 6(3):239–251.
5. Zweigenbaum, P., Demner-Fushman, D., Yu, H., Cohen, K.B. 2007. Frontiers of biomedical text mining: current progress. *Briefings in Bioinformatics.* 8(5):358–375.
6. See http://biocreative.sourceforge.net/
7. Kim, J.-D., Ohta, T., Tsuruoka, Y., Tateisi, Y., Collier, N. 2004. Introduction to the Bio-Entity Recognition Task at JNLPBA. *Proceedings of the International Workshop on Natural Language Processing in Biomedicine and its Applications* (JNLPBA-04), pp. 70–75.
8. Liu, H., Torii, M., Hu, Z.-Z., Wu, C. 2007. Gene mention and gene normalization based on machine learning and online resources. *Proceedings of the Second BioCreative Challenge Evaluation Workshop*, 135–140s.
9. Hanisch, D., Fundel, K., Mevissen, H.T., Zimmer, R., Fluck, J. 2005. ProMiner: rule-based protein and gene entity recognition. *BMC Bioinformatics.* 6(Suppl 1):S14.
10. See http://www.ebi.ac.uk/embl/
11. See http://www.rcsb.org/pdb/home/home.do
12. See http://www.ncbi.nlm.nih.gov/entrez/query/static/help/LL2G.html
13. Xu, H., Fan, J.-W., Hripcsak, G., Mendonça, E.A., Markatou, M., Friedman, C. 2007. Gene symbol disambiguation using knowledge-based profiles. *Bioinformatics*, 23(8): 1015–1022.
14. See www.uniprot.org
15. See http://pubchem.ncbi.nlm.nih.gov/.
16. See http://www.drugbank.ca/
17. See http://www.ebi.ac.uk.chebi/.
18. See http://www.ncbi.nlm.nih.gov/projects/SNP/

19. Klinger, R., Friedrich, C.M., Mevissen, H.T., Fluck, J., Hofmann-Apitius, M., Furlong, L.I., Sanz, F. 2007. Identifying gene-specific variations in biomedical text. *Journal of Bioinformatics and Computational Biology*. 5(6):1277–96.
20. Rebholz-Schuhmann, D., Marcel, S., Albert, S., Tolle, R., Casari, G., Kirsch, H. 2004. Automatic extraction of mutations from Medline and cross-validation with OMIM. *Nucleic Acids Research*. 32(1):135–142.
21. Zhou, G., Zhang, J., Su, J., Shen, D., Tan, C. 2004. Recognizing names in biomedical texts: a machine learning approach. *Bioinformatics*. 20(7):1178–1190.
22. See http://www.nlm.nih.gov/mesh/
23. See http://www.geneontology.org/
24. See http://www.pantherdb.org/
25. See http://www.arabidopsis.org/about/index.jsp
26. Thimm, O., Blasing, O., Gibon, Y., Nagel, A., Meyer, S., Kruger, P., Selbig, J., Muller, L.A., Rhee, S.Y., Stitt, M. 2004. MAPMAN: a user-driven tool to display genomics data sets onto diagrams of metabolic pathways and other biological processes. *Plant Journal*. 37(6):914–939.
27. See http://www.sequenceontology.org/
28. See http://www.gopubmed.org/
29. Doms, A., Schroeder, M. 2005. GoPubMed: exploring PubMed with the Gene Ontology. *Nucleic Acids Research*. 33(Web Server issue):W783–W786.
30. Müller, H.M., Kenny, E.E., Sternberg, P.W. 2004. Textpresso: an ontology-based information retrieval and extraction system for biological literature. *PLoS Biology*. 2(11):e309.
31. See, for example, http://www.gopubmed.org/
32. Lisacek, F., Chichester, C., Kaplan, C., Sandor, A. 2005. Discovering paradigm shift patterns in biomedical abstracts: application to neurodegenerative diseases. *Proceedings of the First International Symposium on Semantic Mining in Biomedicine* (SMBM) 2005, Hinxton, Cambridge, UK. See also http://ftp.informatik.rwth-aachen.de/Publications/CEUR-WS/Vol-148/
33. Dr. Philippe Bijenga, University Hospital Geneva, personal communication
34. See http://sig.biostr.washington.edu/projects/fm/AboutFM.html
35. See http://www.who.int/classifications/icd/en/
36. See http://pubchem.ncbi.nlm.nih.gov/
37. See http://www.ebi.ac.uk/chebi/
38. See http://www.drugbank.ca/
39. See http://www.hmdb.ca/
40. See http://www.infonortics.com/chemical/ch07/slides/thielemann.pdf
41. Lafferty, J., McCallum, A., Pereira, F. 2001. Conditional random fields: probabilistic models for segmenting and labeling sequence data. In *Proceedings of the 18th International Conference on Machine Learning*, San Francisco: Morgan Kaufmann, pp. 282–289.
42. Corbett, P., Murray-Rust, P. 2006. High-throughput identification of chemistry in life science texts. In *Computational Life Sciences II*, Lecture Notes in Computer Science series, Vol. 4216. Berlin/Heidelberg: Springer, pp. 107–118.
43. Kemp, N., Lynch, M. 1998. The extraction of information from the text of chemical patents. 1. Identification of specific chemical names. *Journal of Chemical Information and Computer Sciences*. 38:544–551.
44. Anstein, S., Kremer, G., Reyle, U. 2006. Identifying and classifying terms in the life sciences: the case of chemical terminology. In N. Calzolari, K. Choukri, A. Gangemi, B. Maegaard, J. Mariani, J. Odijk, D. Tapias (Eds.), *Proceedings of the Fifth Language Resources and Evaluation Conference*, pp. 1095–1098.
45. Corbett, P., Batchelor, C., Teufel, S. 2007. Annotation of chemical named entities BioNLP. *Biological, Translational, and Clinical Language Processing*. 57–64.

46. See http://www.infonortics.com/chemical/ch07/slides/hofmann.pdf

47. McDaniel, R., Balmuth, J.R. 1992. Kekule: Ocr-optical chemical (structure) recognition. *Journal of Chemical Information and Computer Sciences.* 32(4):373–378.

48. Ibison, P., Jacquot, M., Kam, F., Neville, A.G., Simpson, R.W., Tonnelier, C., Venczel, T., Johnson A.P. 1993. Chemical literature data extraction: the CLiDE Project. *Journal of Chemical Information Computer Science.* 33(3):338–344.

49. Fluck, J., Zimmermann, M., Kurapkat, G., Hofmann, M. 2005. Information extraction technologies for the life science industry. *Drug Discovery Today-Technologies.* 2(3):217–224.

50. See http://cactus.nci.nih.gov/cgi-bin/osra/index.cgi

51. See http://www.simbiosys.ca/clide/

52. See http://www.scai.fraunhofer.de/fileadmin/download/vortraege/tms_07/Martin_Hofmann-Apitius.pdf

53. See http://en.wikipedia.org/wiki/Natural_language_processing

54. See http://mmtx.nlm.nih.gov/index.shtml

55. See http://www.nlm.nih.gov/research/umls/

56. Rindflesh, T.C., Fiszman, M. 2003. The interaction of domain knowledge and linguistic structure in natural language processing: interpreting hypernymic propositions in biomedical text. *Journal of Biomedical Informatics.* 36:462—477.

57. Greenwood, M.A., Stevenson, M., Guo, Y., Harkema, H., Roberts, A. 2005. Automatically acquiring a linguistically motivated genic interaction extraction system. In J. Cussens, C. Nédellec (Eds.), *Proceedings of the Workshop on Learning Language in Logic (LLL05),* pp. 46–52.

58. Plake, C., Hakenberg, J., Leser, U. 2005. Optimizing syntax patterns for discovering protein-protein interactions. ACM Symposium on Applied Computing (SAC), Bioinformatics Track.

59. Jelier, R., Jenster, G., Dorssers, L.C.J., van der Eijk, C.C., van Mulligen, E.M., Mons, B., Kors, J.A. 2005. Co-occurrence based meta-analysis of scientific texts: retrieving biological relationships between genes. *Bioinformatics.* 21(9):2049–2058

60. Hoffmann, R., Krallinger, M., Andres, E., Tamames, J., Blaschke, C., Valencia, A. 2005. Text mining for metabolic pathways, signaling cascades, and protein networks. *Science STKE.* 2005(283):pe21.

61. Gonzalez, G., Uribe, J.C., Tari, L., Brophy, C., Baral, C. 2007. Mining gene-disease relationships from biomedical literature: weighting protein-protein interactions and connectivity measures. *Pacific Symposium on Biocomputing,* pp. 28–39.

61a. Singh, S.B., Hull, R.D., Fluder, E.M. 2003. Text Influenced Molecular Indexing (TIMI): a literature database mining approach that handles text and chemistry. *Journal of Chemical Information and Computer Sciences,* May-June, 43(3): 743–752.

62. See http://www.aneurist.org

63. Gattermayer, T. 2007. *SCAIView Annotation and Visualization System for Knowledge Discovery,* Bonn-Aachen International Center for Information Technology, University of Bonn, Germany, October 2007.

64. See and Rebholz-Schuhmann, D., Kirsch, H., Arregui, M., Gaudan, S., Riethoven, M., Stoehr, P. 2007. EBIMed—text crunching to gather facts for proteins from Medline. *Bioinformatics.* 23(2):e237–e244.

65. See http://www.nlm.nih.gov/medlineplus/druginformation.html

66. Corbett, P., Murray-Rust, P. 2006. High-throughput identification of chemistry in life science texts. *Springer Lecture Notes in Computer Science,* Vol. 4216, pp. 107–118.

67. Chang, J.T., Altman, R.B. 2004. Extracting and characterizing gene-drug relationships from the literature. *Pharmacogenetics.* 14(9):577–586.

68. Ahlers, C.B., Fiszman, M., Demner-Fushman, D., Lang, F.-M., Rindflesch, T. 2007. Extracting semantic predications from Medline citations for pharmacogenomics. *Pacific Symposium on Biocomputing.* 12:209–220.

69. Kolárik, C., Hofmann-Apitius, M., Zimmermann, M., Fluck, J. 2007. Identification of new drug classification terms in textual resources. *Bioinformatics*. 23 (13):i264–i272.

70. Frijters, R., Verhoeven, S., Alkema, W., van Schaik, R., Polman, J. 2007. Literature-based compound profiling: application to toxicogenomics. *Pharmacogenomics*. 8(11): 1521–1534.

71. See www.research.ibm.com/UIMA/

72. See http://uima-framework.sourceforge.net/ and http://www.julielab.de/content/view/122/179/

73. See http://www.temis.com/?id=32&selt=14

74. See http://www.research.ibm.com/journal/sj43-3.html and publications therein

75. Mack, R., Mukherjea, S., Soffer, A., Uramoto, N., Brown, E., Coden, A., Cooper, J., Inokuchi, A., Iyer, B., Mass, Y., Matsuzawa, H., and Subramaniam, L.V. 2004. Text analytics for life science using the Unstructured Information Management Architecture. *IBM Systems Journal*. 43(3).

76. Shannon, P., Markiel, A., Ozier, O., Baliga, N.S., Wang, J.T., Ramage, D., Amin, N., Schwikowski, B., Ideker, T. 2003. Cytoscape: a software environment for integrated models of biomolecular interaction networks. *Genome Research*. 13(11):2498–2504. See also http://www.cytoscape.org/

77. Plake, C., Schiemann, T., Pankalla, M., Hakenberg, J., Leser, U. 2006. Ali Baba: PubMed as a graph. *Bioinformatics*. 22(19):2444–2445.

78. See http://domino.research.ibm.com/comm/research_projects.nsf/pages/uima.index.html

8 Semantic Web

Colin Batchelor and Richard Kidd

CONTENTS

INTRODUCTION

The *Semantic Web* is a vision of the World Wide Web where the pages can be, in a manner of speaking, understood by computers (Berners-Lee et al. 2001). What this requires is a consistent set of machine-readable identifiers for concepts and a defined set of logical relations that can be used to draw inferences about them and reason over papers. This is to be distinguished from the natural language processing problem of question answering, where queries in natural language, such as "Where are the pubs in Richmond?", are answered by parsing web pages themselves. As we shall see later on, natural language processing is a promising way of bridging between the human-readable and machine-readable web.

The technology of Web services is closely related. To process the information passed to them appropriately, Web services need to know what sort of information is being provided, and ontologies and identifiers can play an important role in this.

The standards body for the World Wide Web as a whole, and hence for the Semantic Web, is the World Wide Web Consortium (W3C; www.w3.org), which is based at ERCIM in Sophia-Antipolis in France, MIT in the United States, and Keio University in Japan.

The possibilities and application of the Semantic Web to chemistry were initially identified by Murray-Rust and Rzepa (1999, 2000; Rzepa and Murray-Rust 2001) and have been promoted in a number of papers since. The precondition to the Semantic Web, the maintenance of data within XML format, has become a reality, whereas the representation of the data within XML has been subject to evolution in the last few years. For example, the SVG format for holding graphics as XML was initially promising, but adoption never took off beyond a few examples. As Adobe is no longer developing and supporting the format, it can be regarded at present as an evolutionary dead end.

In this chapter we shall first describe RDF and OWL, the grammar of the Semantic Web, and then move on to identifiers, the vocabulary of the Semantic Web, and ontologies, which represent the real-world knowledge required to make use of the grammar and vocabulary. We will concentrate on practical chemistry- and biochemistry-orientated deployments of Semantic Web technology rather than the computer science behind it. Then we will cover some case studies: Web services, databases, and semantic publishing in the forms of Semantic Eye and RSC Project Prospect. We will briefly cover what the Semantic Web has to offer for experimental data before finishing up with some possible future directions.

THE SEMANTIC WEB

GRAMMAR

Resource Description Format (RDF) is a way of making statements about subject–object–predicate triples. The subject, object, and predicate of each statement should each, at least in principle, have a Universal Resource Identifier (URI). Much of this chapter will be devoted to describing existing schemes for assigning URIs to subjects and objects. Rather less, and less rigorous, work has been done on predicates.

In a statement such as "caffeine is a diuretic," "caffeine" is the subject, "diuretic" is the object, and "is a" is the predicate. The Semantic Web stands or falls not only on identifiers but also on the consistent and well-documented use of predicates.

RDF itself says nothing about how these triples should be represented. We shall discuss one way of converting these triples into text, or *serializing* them, later on.

An *ontology* is a shared formal representation of knowledge. The strength of an ontology over a database, or a simple controlled vocabulary, is that it contains predicates that enable it to be reasoned over. The word itself, from the classical Greek for *being* and *discourse*, describes the philosophical pursuit, going back to Aristotle's *Categories*, of classifying what is, but has since come to mean a computational artifact for representing real-world knowledge.

Then we can divide ontologies into upper-level ontologies, such as Suggested Upper Merged Ontology (SUMO) (Niles and Pease 2001) or Basic Formal Ontology (BFO) (Grenon et al. 2004), which attempt to provide a framework for describing everything in the universe, and domain ontologies, which are focused on knowledge in a particular area, such as transmissible diseases, zebrafish anatomy, or indeed chemical structure.

It is important that upper-level ontologies do not attempt to be too ambitious in their coverage of domain-specific information. SUMO attempts to cover basic chemistry and gets it badly wrong. Molecule, for example, is defined as, "A molecule is the smallest unit of matter of a CompoundSubstance that retains all the physical and chemical properties of that substance, e.g., Ne, H_2, H_2O." It manages to be inconsistent because it then defines a CompoundSubstance in such a way as to exclude hydrogen molecules from the definition of Molecule. The definition of Metal is not only a sibling of Atom and SubatomicParticle, but excludes alloys. BFO has less ambitious aims than SUMO and is deliberately minimal in its scope.

Traces of the early ontology work are still apparent. The best practice enshrined by the OBO Foundry (Smith et al. 2007) involves a so-called Aristotelian specification of how to write the definitions for the entries in an ontology. The definition should be written in terms of the genus, the "kind" of thing you are describing, and its differentia, what distinguishes it from its siblings. Hence "man is an animal that is rational," or, to take an example from the Sequence Ontology (Eilbeck et al. 2005), a primary transcript (SO:0000185) is "a transcript that in its initial state requires modification to be functional."

It is worth mentioning that ontologies describe types of things, rather than things you would find in the universe, which are instances of the types. A building ontology would have entries for "detached house," "apartment," and "shed," rather than "23 Acacia Avenue, Framley." For chemistry, a type would be "water," and its instance would be a particular molecule of water in the sea or in a cup of coffee.

The two foundational relationships between entries in an ontology are is_a and part_of. The is_a relationship is a taxonomic one; thus, a dog is_a animal, a spaniel is_a dog, and so forth. Likewise, the part_of relationship specifies that a hinge is part_of a door, a door is part_of a room, and so on. However, there is a subtlety. Not all parts can be described with the part_of relationship, because if you say that type A part_of type B, that means that all instances of A are necessarily part of some instance of B. So hydrogen is not part_of methane. There are very few genuine cases of a part_of relationship that describes chemical compounds. It is better to use a reciprocal relationship, has_part, in which *all* type A has_part *some* type B, or all methane molecules have some hydrogen atoms as parts.

RDF Schema (RDFS) (Brickley and Guha 2004) is an extension of RDF to cover simple ontologies. It adds classes and resources and properties, such as rdfs:subClassOf, which allow arbitrary relationships to be defined, or rdfs:range and rdfs:domain, which allow the scope of relationships to be restricted. rdfs:label provides a human-readable name as an alternative to the URI, and rdfs:comment can be used for human-readable definitions. Web Ontology language (OWL) (McGuinness and Van Harmelen 2004) is a way of representing important ontological concepts such as disjointness that are not covered in RDF or RDFS, but it is outside the scope of this chapter.

VOCABULARY

Standardization of terminology is the first problem to be solved, and in chemistry this can be broken down into several areas: substance identification, reaction and experimental information, subject terminology, and relationships between them. In some cases this is a case of unique identifiers (in the case of substances), in some having an agreed set of terms and relationships (such as reaction types) and finally of having structured ways of describing experimental and property data.

A famous zoological classification divides animals into the following:

1. Those that belong to the Emperor
2. Embalmed ones
3. Those that are trained
4. Suckling pigs
5. Mermaids
6. Fabulous ones
7. Stray dogs
8. Those included in the present classification
9. Those that tremble as if they were mad
10. Innumerable ones
11. Those drawn with a very fine camelhair brush
12. Others
13. Those that have just broken a flower vase
14. Those that from a long way off look like flies

This fanciful classification, attributed to a Chinese encyclopedia in Borges's short story "The Analytical Language of John Wilkins," and brought to the world's attention by Foucault in his *The Order of Things* is clearly ludicrous, but perhaps not so different from the root classification of chemical compounds in the Medical Subject Headings (MeSH), which is as follows:

1. Inorganic chemicals
2. Organic chemicals
3. Heterocyclic compounds
4. Polycyclic compounds
5. Macromolecular substances
6. Hormones, hormone substitutes, and hormone antagonists
7. Enzymes and coenzymes
8. Carbohydrates
9. Lipids
10. Amino acids, peptides, and proteins
11. Nucleic acids, nucleotides, and nucleosides
12. Complex mixtures
13. Biological factors
14. Biomedical and dental materials
15. Pharmaceutical preparations
16. Chemical actions and uses

Many, if not most, compounds of medical interest fall into many of the above categories, just as many if not most animals fall into many of Borges's. Properly constructed ontologies obey the principle of *pairwise disjointness*, that is, no term can belong to two sibling categories. One risk of not following this principle is that the ontology can end up containing cycles, which would cause a reasoner, a computer program that carries out inference over an ontology, to enter an infinite loop.

The Crystallographic Information Format (CIF) developed by the International Union of Crystallography (IUCr) (Hall et al. 1991) is the best applied standard within chemistry and is an exemplar of the difficulties of standardization. The CIF is well accepted and understood by the community, is supported by publishers, and easy-to-use validation tools exist. However, ensuring that the resulting CIFs are valid and accurate requires additional quality assurance; publishers of CIFs may carry this out (as, for example, the RSC does). So the development and adoption of standards is not sufficient to ensure that accurate and unambiguous data will be presented within the standards.

In this section we will describe the three general identifier strategies for objects of chemical discourse: InChIs (already introduced in Chapter 3), ontology IDs, and BioRDF.

The important and classifiable objects of chemical discourse include, but are not restricted to:

1. Compounds of known structure
2. Compounds of unknown structure
3. Mixtures of compounds
4. Classes of compound
5. Parts of molecules
6. Chemical elements in the pure state
7. Chemical elements as measured by analytical chemistry techniques
8. Biological sequences of naturally occurring nucleotides or peptides
9. Gene products, that is, proteins and RNAs
10. Organisms

Members of classes 1 to 7 are strictly defined by their constituent parts; members of classes 9 and 10, loosely so, so that they can lose parts and retain their identity.

Out of all of the above, only the InChI can unambiguously describe some compounds of known structure. The rest have a variety of representations, many of which are keyed to entries in an ontology. BioRDF is a sort of wrapper for database IDs that describe biological entities.

Chemical Compounds and the Semantic Web

InChI strings can identify chemical compounds, class 1 in our classification. The drawbacks to CAS numbers and Simplified Molecular Line Entry System (SMILES) have been rehearsed elsewhere, so we shall briefly mention that they are unsuitable for URI-fication. The URI for an InChI, as defined by the INFO registry, which is run by the National Information Standards Organization in the United States, is made by prefixing the InChI with "info:inchi=" and URI-encoding contentious characters such as plus signs. The unique property of the InChI URI, as opposed to others, such as digital object identifiers (DOIs) or PubMed IDs, is that it contains all of the information needed to reconstruct the chemical structure by referring only to an algorithm, rather than an external database. However, it is potentially long and unwieldy, so the InChIKey has been developed. This is a hashed, fixed-length version of the InChI. Unfortunately, it is impossible to reverse the hashing process and extract the original structure from the InChIKey.

It is not necessarily clear whether a given chemical name indicates something that belongs to class 1, class 4, or class 5. To illustrate the difficulty in determining whether a given chemical name refers to a compound, a class of compounds, or a part of a molecule, and how we would represent it if it did not represent a compound, let us consider the word *imidazole*. It can stand for a chemical compound, which can be described by an InChI. However, this is rare. Its two main uses are to indicate membership of a class of compounds or a part of a molecule. Generally, if a chemical compound name has a determiner, or is plural, for example, "the imidazole **10**," where **10** is the number of a structure in a figure, or "The imidazoles are. . . ," the name will not refer to imidazole itself, but to a compound that contains the imidazole skeleton. Similarly, phrases like "the imidazole sidechain" refer to part of the amino acid histidine, which is distinguished from the other amino acids by containing the imidazole skeleton. The histidine in question is unlikely to be a free molecule, but would rather be an amino acid residue and itself part of a much larger system.

That a term in ChEBI refers to a class (our class 4) of compounds, if it is a class defined by structural characteristics, is indicated by its name being plural. There is no straightforward distinction in ChEBI between classes of which the instances are other classes and classes of which the instances are molecules, atoms, and so on.

Parts (our class 5) are more problematic. Only those parts that are substituents (in ChEBI called "groups") have entries in ChEBI. These are typically clearly linguistically signaled in text—imidazolyl, pyrrolyl, methyl, and so forth. Those parts onto which the groups are substituted—benzene rings, cyclohexanone skeletons, and so forth—do not exist in ChEBI.

We have no framework for representing compounds of unknown structure (class 2 in the above classification), such as you will find discussed in natural products chemistry, other than the name, but the name is not itself an unambiguous identifier.

It might be thought that element names stand unambiguously for the elements themselves, mapping onto an InChI of the sort InChI=1/Au. However, in analytical chemistry and metallomics, the phrase "determination of copper" tells us little about the chemical environment of the copper atoms, whether they are bonded, and so forth. ChEBI does distinguish between "copper" (our class 7) and "elemental copper" (our class 6), but this is poorly documented.

Ontology IDs: The Case of OBO

The OBO format for describing ontologies developed out of work on the Gene Ontology (Gene Ontology Consortium 2000) independently from the W3C's work on OWL, the aim being to have a lightweight, human-readable, human-writable text-based format. Moreira and Musen (2007) developed a non-lossy way of converting OBO to OWL and vice versa, and with it necessarily a URI-fication of OBO identifiers. They look like this:

```
http://purl.org/obo/owl/CHEBI:27899
```

which is the ChEBI reference for the square-planar complex cisplatin $[Pt(Cl)_2 (NH_3)_2]^{2+}$, where like ligands are adjacent to one another. Transplatin, also $[Pt(Cl)_2 (NH_3)_2]^{2+}$, except that here unlike ligands are adjacent to one another, by contrast, is

```
http://purl.org/obo/owl/CHEBI:35852
```

The two molecules share the same InChI because it cannot represent non-sp^2 or sp^3 coordination environments, unlike SMILES.

As for our class 8, there is some ambiguity for very short biological sequences. Given an appropriately capped terpeptide GlyLysSer, the representations GlyLysSer, GLS, or the relevant InChI will all be equally meaningful, but there is no fundamental representation, given the limits on the length of an InChI, that could cope with a chiliapeptide (with 1,000 bases).

The Gene Ontology (Gene Ontology Consortium 2000) is not an ontology of genes, nor of gene products (our class 9), that is, proteins and RNAs, especially noncoding RNAs. What it does provide identifiers for, are the functions of a gene product, the biological processes it may be involved in, and the locations in a cell where these gene products may act. These cellular components are sometimes themselves gene products, but we shall not consider them further here.

BioRDF and Other Biological Identifiers

BioRDF (Ruttenberg et al. 2007) supports a variety of neurochemical databases, has RDF-ified gene records, databases of receptor–ligand and protein–protein interactions, a directory of commercially available antibodies, Reactome, KEGG, the NCI metathesaurus, and UniProt (which is a collaboration between the SIB in Zurich and the EBI in Hinxton, UK), hence providing a Semantic Web representation of protein sequences.

However, gene products in general are trickier. We not only have to consider bare protein sequences, but also their three-dimensional structures, which are stored in the PDB and their family data in Pfam. All of these and many more sources are intended to be pulled together by the Protein Ontology (PRO) project (Natale et al. 2006), which has one component based on evolutionary relatedness and another component based on protein structure. The parallel RNA Ontology (RNAO) project (Leontis et al. 2006) is intended to do something similar for RNAs and their structures but does not cover family data, which is held by RFAM (Griffiths-Jones et al. 2005).

As for organisms, our class 10, Linnaeus had this sorted out in the 18th century with his *Systema Naturae*. With some modifications (these days you and I are *H. sapiens* rather than *H. diurnus*—*H. nocturnus* was the orangutan!), his binomial system has survived to the present day, and of course is eminently well suited to being mined out with a simple regular-expression-based system. The options for (slightly) more machine-readable identifiers are the NCBI Taxonomy and the Life Science Identifier (LSID) project (see the LSID resolver at http://lsid.tdwg.org/).

CASE STUDIES

The main application areas of Semantic Web technology to chemistry so far have been in Web services, especially PubChem, chemical databases, and publishing.

Although there are a number of examples of the application of Semantic Web technologies to scientific publishing, for example, Ingenta translating parts of their publications database into RDF and keeping it in a triplestore (Portwin and Parvatikar 2006), we shall concentrate on those relevant to chemistry.

WEB SERVICES

A Web service is a computer program that is accessible over the Web. There are two architectures for doing this: REST (Fielding 2000) and SOAP (Mitra and Lafon 2007). Each SOAP Web service is described in terms of Web Service Description Language (Booth and Liu 2007), which is a specification in XML. This enables data passed over the Web, say a sequence of characters such as CMRSGGCTRRYAC, to have its type specified according to an ontology, thus telling the code that that sequence of characters is a consensus DNA sequence rather than an author name or a geographical location. A "RESTful" Web service works in terms of HTTP requests and thus has potentially a less constrained syntax.

PubChem provides examples of both sorts of Web service. To search PubChem for a particular chemical name, then, you can use the Entrez Programming Utilities (http://www.ncbi.nlm.nih.gov/books/bv.fcgi?rid=coursework.chapter.eutils). Requests to these go via the URL and can be typed into a browser by hand. For more sophisticated requests, PubChem provides the Power User Gateway (http://pubchem.ncbi.nlm.nih.gov/pug/pughelp.html). This uses a complicated XML format for requests, much of which is there for job control for large batches.

One important application of Web services is in distributing computational effort over many different machines, creating what in the UK is called a grid, or in the cases we are about to mention, a semantic grid. We shall briefly describe two research projects in the UK that have applied this to the field of chemistry, in Southampton and Leeds, respectively.

The UK National Crystallography Service (NCS) in Southampton has an e-science infrastructure for conducting small-molecule crystallography experiments and disseminating them over the Web. This has developed from the Combechem project, which has also focused on capturing the entire experimental process with, for example, semantically rich electronic lab books (Hughes et al. 2004). In their description of the eCrystals server, Coles et al. (2006) describe in detail how to authenticate users, control experimental apparatus, and organize job queues. However, what is most relevant to the Semantic Web is how the crystal data obtained is disseminated. Bibliographic metadata aside, the points of chemical interest are the chemical formula, InChI (as we have seen earlier), compound class, crystallographic keywords, and the stages of the experiment for which data files are present. These are disseminated through the Open Archives Initiative via qualified Dublin Core terms (Hillmann 2005). The author describes the terms currently used (Koch and Duke 2006) as placeholders "until official ones become available."

Similar in spirit, but aimed at a very different community, is the work at Leeds, carried out jointly between the research groups of Peter Dew in computing and Mike Pilling in chemistry. One important focus here is on distributing the computational task of modelling chemical reaction processes in reaction kinetics, combustion, and atmospheric chemistry (Pham et al. 2005, 2006). The other focus, and this is more of a desideratum than something that has been fully accomplished, is getting hold of useful reaction data, mechanisms, and models, which are often scattered between research groups and not necessarily published.

DATABASES: CHEMBLAST

Bhat and Barkley (2007) provide the most impressive case yet of integrating disparate databases through the Semantic Web, these being the Protein Databank, the HIV structural database (AIDSDB), and PubChem. Their aim is to create a database of AIDS inhibitor drugs and how these drugs bind to active sites of enzymes. They achieve this by a URI-fication of the InChI (Prasanna et al. 2005) both for whole molecules and, rather controversially, for parts of molecules, and a chemical taxonomy (Prasanna et al. 2006) that describes whole molecules, their constituent groups, and the "fragments" that might be used in structure searching with a carefully described, ontologically well-formed set of relationships between them.

A particularly innovative feature of this work is that it treats the InChI of the drug molecule as an invariant URI and provides rules for generating local, application-specific URIs (in their terms, semi-invariant or ontologically defined URIs: OURIs) for the different needs of, for example, molecular modelers, medicinal chemists, or biologists.

Their URI-fication of the InChI, and the URI-fication by the INFO registry, are not identical, so there is some standardization yet to be achieved. Likewise, their chemical taxonomy and ChEBI are not yet interoperable.

PUBLISHING: SEMANTIC EYE

Semantic Eye (Casher and Rzepa 2006) is a test-of-principle scheme for the semantic enrichment of journal articles. Rather unusually, it treats the PDF as the locus of semantic enrichment. The identifiers are mapped onto RDF triples that are then serialized as XML using Adobe's Extensible Metadata Platform (XMP) schema within the PDF. It uses InChIs as identifiers for molecules and DOIs for the articles themselves. The idea is that the identifiers derived from the PDFs can be stored locally on a user's machine inside the PDFs, which are then mined by desktop indexing services, creating a sort of semantic intranet or semantic desktop. Exactly how the identifiers are assigned to the papers in the first place is left open to the user.

PUBLISHING: RSC PROJECT PROSPECT

History and Development Route

Project Prospect is at the time of writing the first real application of semantic enhancement to primary research literature. By using open standards such as the InChI and the Open Biomedical Ontologies, the aim was to remove the ambiguity of searching (this remains to be well integrated with the search engines), but the information is now held in a structured form that can make this happen. In this first implementation, the information is used to add a layer of additional information (visualizations and definitions) and identify relationships between our own related HTML articles.

Along with other publishers, RSC has sponsored summer students at the Unilever Centre of Molecular Informatics at Cambridge University (within Peter Murray-Rust's group) for a number of years. Out of these projects evolved a number of software tools, such as the Experimental Data Checker (Adams et al. 2004), and a greater understanding of the possibilities of using structured data within the publication process. The RSC internal development project started about a year before launch and had a number of aims: to use open standards for subject and chemical terms to allow better identification of relevant content by search engines and, consequently, readers, to develop the display and reuse of structured experimental data within publication workflow, and to apply this across the RSC's published content. The development of the IUPAC InChI identifier as an open standard for representing a chemical substance also enabled many compounds to be dealt with in a sensible manner. The opportunities offered led RSC to set up this Project Prospect to both implement them in a sustainable form into the journal production workflows and to demonstrate the possibilities to readers. The production route uses the OSCAR3 text mining package (part of the SciBorg [Copestake et al. 2006] collaboration between the Unilever Centre for Molecular Informatics and the Computer Laboratory, both at the University of Cambridge) to identify the compounds and subject terms, open standards to keep the metadata within XML, and amendments to article publishing routines to display the data online.

Semantic Content

Scalable assignment of identifiers to papers was one of the basic design criteria behind Project Prospect, which is significantly more ambitious but as yet only semi-detached part of the Semantic Web. Here the RDF is embedded in the RSS feeds. Currently, the identifiers in use are InChIs for InChI-fiable chemical compounds and OBO IDs for terms found in the Gene Ontology (Gene Ontology Consortium 2000), Sequence Ontology (Eilbeck et al. 2005), and the OBO cell-type ontology (Bard et al. 2005). The identifiers are assigned to the articles by text mining and manual curation.

It will be useful here to show how the RDF of the RSS feeds are serialized as XML. This pulls together web feeds, metadata for publishing and chemistry, and human- and machine-readable information. Some knowledge of XML will be needed to understand what follows.

The first thing to note is that in RSS 1.0, everything is wrapped up in an <rdf:RDF> element that defines the XML namespaces used throughout the document. Inside this is a <channel> element. This is the subject of the first set of relationships. It is identified by the rdf:about attribute.

The XML that defines the semantic content of each paper works like this: Each article is represented by an <item> element with an rdf:about attribute, indicating that the article is the subject of an RDF triple. It could be defined with a doi: URI, a pmid: URI (for those RSC articles that are abstracted by PubMed) or, as it is in practice, by a URL that points to the RSC's DOI resolver. A <content:items> element, as defined by the "content" module of RSS 1.0, has within it an <rdf:Bag> element that contains <rdf:li> elements, and within each of those a <content:item> element whose rdf:about attribute is the URI of the entity mentioned within the article. This means <content:item> is the predicate, or the verb, of the relationship. Written without the XML, a relationship might look like this:

```
doi:10.1039/b715455k content:item http://purl.org/obo/
owl/SO#SO:0001078
```

which in English says

```
(Chiovitti et al. 2008) mentions polypeptide secondary
structure.
```

The publishing content is specified using the Dublin Core and PRISM specifications. The <dc:publisher> element, for example, specifies an as_publisher relationship between the article and the human-readable text "The Royal Society of Chemistry." In time we may expect that this text could be replaced by a URI for the publisher.

The important question for how well this works as part of the Semantic Web is how well the resolution mechanisms work. IUPAC provides no resolution mechanism for InChI URIs. The OBO URIs themselves resolve to the ontologies themselves represented in OWL. The RSC's implementation has to be considered incomplete because a DOI does not yet resolve to a page containing machine-readable information about the paper. Only when this is achieved can Project Prospect be properly considered part of the Semantic Web.

Results and Applications

Currently, the included metadata are used to create additional functionality for the reader within an enhanced HTML view of an article. The ontology terms link to pop-up pages with the ontology definitions, further links, and related articles, whereas the compounds bring up a pop-up containing a two-dimensional structure, the InChI, and SMILES strings for the compound, names, synonyms, and related articles. This is best shown in Figure 8.1.

The enhanced RSS feeds, described above in the section about serialization, are a unique innovation and go beyond the bibliographic information and graphical abstracts that are now standard. They are also open to all readers, so anyone, subscriber or not, can put the RSC feeds straight into a database and get a very good idea of the compounds within a newly published article and, to some extent, their biological activity.

Project Prospect was launched in February 2007 and was the winner of the 2007 ALPSP/Charlesworth Award for Publishing Innovation, which recognizes a significantly innovative approach to any aspect of scholarly publication. The judging panel considered that Project Prospect "was the clear winner . . . with an elegant and intuitive on screen manifestation of the advantages of including . . . metadata. As a result, sophisticated and effective searching of the literature is greatly improved and the value gained from reading each article is significantly enhanced. *Project Prospect* is delightfully simple to use and its benefits to authors and readers are immediately obvious."

Although the additional metadata is not being significantly picked up by the search engines, and this currently is a project restricted to RSC publications, it can be thought of as a contribution to the development of the Semantic Web rather than a real part of it. However, its development does give an excellent object lesson in

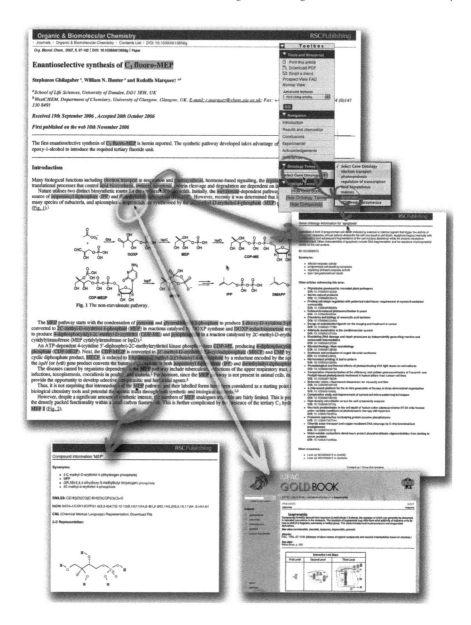

FIGURE 8.1 *A color version of this figure follows page 146.* Semantic mark-up of content linking to additional data sources.

the types of information that can usefully be identified within chemical papers and the current limitations, but also that this information can provide real and useful enhancements to the accessibility of the published science. When the reader starts to expect this enhancement, and even takes this for granted like they do now with full-text reference linking via CrossRef, this will be a step change in the way our science is published and accessed.

EXPERIMENTAL DATA STANDARDS AND THE SEMANTIC WEB

Chemistry is some way behind the biomedical sciences in standardization of research protocols. The MIBBI (Minimum Information for Biological and Biomedical Investigations) Project (MIBBI Consortium 2008) aims to bring the various Minimum Information protocols for different sorts of biomedical experiments into line with one another and oversees over a dozen protocols. The mapping of the protocols themselves to stable Semantic Web identifiers is achieved through an ontology.

The best example within chemistry so far has been the Crystallographic Interchange Format (CIF) (Hall et al. 1991) developed by the International Union of Crystallography as described above. No such similar standard exists for other machine-based techniques such as NMR, with the different instrument manufacturers having their own mutually incompatible ways of specifying pulse sequences and experimental conditions.

In both inorganic and organic synthetic chemistry, experimental data (elemental composition, melting points, nuclear magnetic resonance, infrared, ultraviolet spectra, and chromatograms) for synthesized molecules are summarized in a consistent and formulaic way. This can be identified and parsed with a simple (although brittle) finite-state model, and the original spectra and chromatograms can be checked for internal consistency and to some extent reconstructed (Adams et al. 2004).

For the actual specification of experiments themselves, two ontologies have grown up independently, one bottom-up and the other top-down. The Ontology for Biomedical Investigation (OBI) is a generalization of the Microarray and Gene Expression Data (MGED) ontology, which was developed to represent microarray experiments, to biomedical experiments in general. Conversely, EXPO (Soldatova and King 2006) was developed from a philosophy-of-science perspective and has been specialized to specific domains, for example, to support the yeast genetics experiments carried out by the Robot Scientist (King et al. 2004; Soldatova et al. 2006). An ongoing project at the University of Wales Aberystwyth is applying EXPO to physical chemistry, and we expect the first results from this project in the near future.

As this chapter is being written, Microsoft has announced a two-year eChemistry project, based on the application of OAI-ORE (Open Archives Initiative–Object Reuse and Exchange, http://www.openarchives.org/ore/) protocols for sharing scholarly information over the Web. Whereas the eChemistry project aims to use these to search and index with specific existing online databases and print archives and work out how best to record chemistry data captured in lab experiments, the development and adoption of new standards for experimental data offer an accessible future for this data that currently is not generally available for readers to find and reuse. Additionally, there is also funding to develop Word 2007 tools for authors to annotate their own papers with ontology and controlled vocabulary information (Bourne and Fink 2007). The authors hope and expect that the acceptance of new standards will offer real benefits to all creators, publishers, and users of chemical information, and the availability of structured experimental data will provide a means of keeping more of the laboratory data within the publishing and discovery workflow.

FUTURE DIRECTIONS

Restricting ourselves to the short-term future, where will the Semantic Web be in two years with respect to chemistry?

Implementing semantic identification is only the first step; the second is the retrieval of this information, which allows the interpretation of the relevance of the subject matter in semantic terms. Although bespoke systems can be used to do this, we really need web-wide search engines to be able to do this interpretation and use the information returned from data sources to generate hit results based on the semantic content rather than full text.

As noted above, the length and composition of the InChI string makes it difficult for search engines to deal with it appropriately, so the fixed-length hashed InChIKey has been developed. InChIs and InChIKeys have been identifiable via Google (with mixed success; see Coles et al. 2005), but currently the common search engines are highly developed for text searching and not developed to recognize identifiers, ascribe meaning to them, and use this meaning to filter results. Although the developers of some search engines have shown interest in developments in this area, it is as yet uncertain how any development arc can apply to chemical data while standards and practice are still evolving rapidly.

Application of accurate subject metadata has long been a goal of publishers and information users, and it continues to be difficult; developments in natural language processing can be expected to increase the accuracy of classifications and chemical entity recognition after publication, but it is impossible in the short term to see that all contextual relevancy can be identified by automated means, avoiding contextually incorrect markup. This effectively leaves the task to be carried out by authors or publishers of the data, unless some degree of inaccuracy proves to be acceptable. Authors are in theory the best placed to carry out and review automated markup, but this places an additional burden on the authoring process, and consistency is uncertain. We know as publishers that authors' use of a fixed Word template can be extremely variable, so it is possible that the addition of further sophisticated markup will only be willingly carried out by the most dedicated.

InChI is a promising start for a chemical identifier, but it can only handle, and will only be able to handle in the near future, that slice of chemistry that is of interest to the pharmaceutical industry. The next revision is expected to handle polymers, but representing spin states (for gas-phase chemistry), different coordination environments, and non–Pauling-type bonding (for inorganic chemistry), and indeed ring–chain tautomerism or axial chirality, are years away. However, despite these limitations, it has been adopted by an increasing number of services. We expect that the burden of representation will be split between refinements to InChI and the chemical ontologies. ChEBI (Degtyarenko et al. 2008) is also promising, but difficult to use for inference, as it does not follow the ontological best practice set out in the OBO Relations Ontology (Smith et al. 2005). We hope that the standardization efforts such as the OBO Foundry (Smith et al. 2007) will result in an ontology that can play a full role in the Semantic Web in future.

Moving beyond the basics, it is also not yet possible to represent the role of a particular entity within the scientific argumentation. A representation such as the one in Project

Prospect can only distinguish degrees of relevance by inclusion or omission, and a more subtle vocabulary is required. Two approaches suggest themselves. The first would be a URI-fication of the Argumentative Zoning (AZ) approach of (Teufel and Moens 2002), which classifies parts of a paper according to whether they discuss work done by the authors or preceding work, and the rhetorical role of the part, whether it describes generally known background information, other people's work described neutrally, contrastively, or as the basis for the current work, and within the current work, statements of the aim of the paper, or simply neutral descriptions of the new work. The second, complementary, approach would be to assign entities to the experimental classes in EXPO (Soldatova and King 2006), though these do not necessarily cover the whole article.

As for representing scientific articles as a whole purely in terms of OWL, this is more complicated than might be supposed. Attentive readers of scientific texts will notice that there is a subtle and comprehensive system for making assertions in terms of possibilities and necessities rather than the simple subject–object–predicate triple of RDF. Authors use phrases such as "It is not impossible that" or "We can assume that" to distance themselves from absolute certainty and to weaken or strengthen their statements.

The impact of the Programmable Search Engine being developed by Google, described in several talks by Steve Arnold (2007), is obviously difficult to guess. In one interpretation it will be sophisticated enough to develop its own ontologies and classify corpora on the fly, whereas in another it will allow Google to sensibly use additional metadata supplied by the webmaster. So in one scenario the classification effort by the webmaster is unnecessary or irrelevant, and in the other it is crucial. However, it is difficult to see that effort spent classifying material on publication will be wasted, and although the Semantic Web required just too much effort for this to be widely applied across the Web as a whole, there are enough signs that use of InChI identifiers, the development of more widely used and accepted chemical ontologies, and the development of some of other experimental data standards offer a reasonable expectation that the Semantic Web ideals will grow, at first in small pockets. As ever, we will see whether the increase in the ability to find information when the semantic enrichment takes place is enough of an evolutionary advantage to change the chemical information ecosystem.

REFERENCES

S. E. Adams, J. M. Goodman, R. J. Kidd, A. D. McNaught, P. Murray-Rust, F. R. Norton, J. A. Townsend, and C. A. Waudby. 2004. Experimental data checker: better information for organic chemists, *Org. Biomol. Chem.,* 2:3067–3070.

S. E. Arnold. 2007. http://www.arnoldit.com/speeches/arnold-icic-2007.pdf.

J. Bard, S. Y. Rhee, and M. Ashburner. 2005. An ontology for cell types, *Genome Biology* 6:R21.

T. Berners-Lee, J. Hendler, and O. Lassila. 2001. The semantic web, *Scientific American,* May 2001, pp. 34–43.

T. N. Bhat and J. Barkley. 2007. Semantic web for the life sciences: hype, why, how and use case for AIDS inhibitors, in *Proceedings of 2007 IEEE Congress on Services*, Salt Lake City, UT, July 9–13, pp. 87–91.

D. Booth and C. K. Liu. 2007. Web Services Description Language (WSDL) Version 2.0 Part 0: Primer, W3C Recommendation June 26. Accessed February 11, 2008. http://www.w3.org/TR/wsdl20-primer/.

P. Bourne and L. Fink. 2007. Reinventing scholarly communication for the electronic age, *CTWatch Quarterly*, 3(3).

D. Brickley and R. V. Guha. 2004. RDF Vocabulary Description Language 1.0: RDF Schema, W3C Recommendation February 10. Available at http://www.w3.org/TR/rdf-schema/. Accessed February 11, 2008.

O. Casher and H. S. Rzepa. 2006. SemanticEye: a Semantic Web application to rationalize and enhance chemical electronic publishing, *J. Chem. Inf. Model.*, 46:2396–2411.

A. Chiovitti, P. Heraud, T. M. Dugdale, O. M. Hodson, R. C. A. Curtain, R. R. Dagastine, B. R. Wood, and R. Wetherbee. 2008. Divalent cations stabilize the aggregation of sulfated glycoproteins in the adhesive nanofibers of the biofouling diatom Toxarium undulatum, *Soft Matter*, doi:10.1039/b715455k.

S. J. Coles, N. E. Day, P. Murray-Rust, H. S. Rzepa, and Y. Zhang. 2005. Enhancement of the chemical Semantic Web through the use of InChI identifiers, *Org. Biomol. Chem.*, 3:1832–1834.

S. J. Coles, J. G. Frey, M. B. Hursthouse, M. E. Light, A. J. Milsted, L. A. Carr, D. DeRoure, C. J. Gutteridge, H. R. Mills, K. E. Meacham, M. Surridge, E. Lyon, R. Heery, M. Duke, and M. Day. 2006. An e-science environment for service crystallography: from submission to dissemination, *J. Chem. Inf. Model.*, 46(3):1006–1016.

A. Copestake, P. Corbett, P. Murray-Rust, C. J. Rupp, A. Siddharthan, S. Teufel, and B. Waldron. 2006. An architecture for language processing for scientific texts, in *Proceedings of the UK e-Science Programme All Hands Meeting 2006 (AHM2006)*, Nottingham, UK.

K. Degtyarenko, P. de Matos, M. Ennis, J. Hastings, M. Zbinden, A. McNaught, R. Alcántara, M. Darsow, M. Guedj, and M. Ashburner. 2008. ChEBI: a database and ontology for chemical entities of biological interest, *Nucleic Acids Res.*, 36(database issue):D344–D350.

K. Eilbeck, S. E. Lewis, C. J. Mungall, M. Yandell, L. Stein, R. Durbin, and M. Ashburner. 2005. The Sequence Ontology: a tool for the unification of genome annotations, *Genome Biology*, 6:R44.

R. T. Fielding. 2000. Architectural Styles and the Design of Network-based Software Architectures, PhD thesis, University of California, Irvine.

The Gene Ontology Consortium. 2000. Gene Ontology: tool for the unification of biology, *Nature Genetics*, 25:25–29.

P. Grenon, B. Smith, and L. Goldberg. 2004. Biodynamic ontology: applying bfo in the biomedical domain, in D. M. Pisanelli (Ed.), *Ontologies in Medicine*, Amsterdam: IOS Press, pp. 20–38.

S. Griffiths-Jones, S. Moxon, M. Marshall, A. Khanna, S. R. Eddy, and A. Bateman. 2005. Rfam: annotating non-coding RNAs in complete genomes, *Nucleic Acids Res.*, 33:D121–D124.

S. R. Hall, F. H. Allen, and I. D. Brown. 1991. The Crystallographic Information File (CIF): a new standard archive file for crystallography, *Acta Cryst.*, A47:655–685.

D. Hillmann. 2005. Using Dublin Core, Dublin Core Metadata Initiative. Available at http://dublincore.org/documents/usageguide/. Accessed February 12, 2008.

G. L. Holliday, G. J. Bartlett, D. E. Almonacid, N. M. O'Boyle, P. Murray-Rust, J. M. Thornton, and J. B. O. Mitchell. 2005. MACiE: a database of enzyme reaction mechanisms, *Bioinformatics*, 21(23):4315–4316.

G. Hughes, H. Mills, D. De Roure, J. G. Frey, L. Moreau, M. C. schraefel, G. Smith, and E. Zaluska. 2004. The semantic smart laboratory: a system for supporting the chemical escientist, *Org. Biomol. Chem.*, 2:3284–3293.

R. D. King, K. E. Whelan, F. M. Jones, P. G. K. Reiser, C. H. Bryant, S. Muggleton, D. B. Kell, and S. G. Oliver. 2004. Functional genomic hypothesis generation and experimentation by a robot scientist, *Nature*, 427:247–252.

T. Koch and M. Duke. 2006. eBank UK: Metadata Terms, UKOLN, University of Bath, UK. Available at http://www.ukoln.ac.uk/projects/ebank-uk/schemas/terms/2006/02/09.

N. B. Leontis, R. B. Altman, H. M. Berman, S. E. Brenner, J. W. Brown, D. R. Engelke, S. C. Harvey, S. R. Holbrook, F. Jossinet, S. E. Lewis, F. Major, D. H. Mathews, J. S. Richardson, J. R. Williamson, and E. Westhof. 2006. The RNA Ontology Consortium: an open invitation to the RNA community, *RNA*, 12:533–541.

D. L. McGuinness and F. van Harmelen. 2004. OWL Web Ontology Language overview, W3C Recommendation, February 10. Available at http://www.w3.org/TR/owl-features/. Accessed 2008-02-11.

MIBBI Consortium. 2008. http://www.nature.com/nbt/consult/index.html, accessed 2008-02-06.

N. Mitra and Y. Lafon. 2007. SOAP Version 1.2 Part 0: Primer (second ed.), W3C Recommendation, April 27, 2007. Available at http://www.w3.org/TR/soap12-part0/. Accessed 2008-02-12.

D. A. Moreira and M. A. Musen. 2007. OBO to OWL: a protege OWL tab to read/save OBO ontologies, *Bioinformatics*, 23(14):1868–1870.

P. Murray-Rust and H. S. Rzepa. 1999. Chemical markup, XML, and the Worldwide Web. 1. Basic Principles, *J. Chem. Inf. Comp. Sci.*, 39:928.

P. Murray-Rust, H. S. Rzepa, M. J. Williamson, and E. L. Willighagen. 2004. Chemical markup, XML and the Worldwide Web. 5. Applications of chemical metadata in RSS aggregators, *J. Chem. Inf. Comp. Sci.*, 44:462–469.

P. Murray-Rust, H. S. Rzepa, M. Wright, and S. Zara. 2000. A Universal approach to Web-based Chemistry using XML and CML, *Chem. Commun.*, 1471.

D. A. Natale, C. N. Arighi, W. C. Barker, J. Blake, T.-C. Chang, Z. Hu, H. Liu, B. Smith, and C. H. Wu. 2007. Framework for a protein ontology, *BMC Bioinformatics*, 8(Suppl 9):S1.

I. Niles and A. Pease. 2001. Towards a standard upper ontology, in C. Welty and B. Smith (Eds.), *Proceedings of the 2nd International Conference on Formal Ontology in Information Systems (FOIS-2001)*, Ogunquit, Maine, October 17–19.

T. V. Pham, P. M. Dew, L. M. S. Lau, and M. J. Pilling. 2006. Enabling e-research in combustion research community, in *2nd IEEE International Conference on e-Science and Grid Computing (e-Science'06)*, IEEE Computer Society Press.

T. V. Pham, L. M. S. Lau, P. M. Dew, and M. J. Pilling. 2005. A collaborative e-Science architecture towards a virtual research environment, in *Challenges of Large Applications in Distributed Environments, 2005. CLADE 2005. Proceedings*. EPSRC, pp. 13–22.

K. Portwin and P. Parvatikar. 2006. Building and managing a massive triple store: an experience report, in *X-Tech 2006*: *Building Web 2.0*, 16–19 May, 2006, Amsterdam.

M. Prasanna, J. Vondrasek, A. Wlodawer ,and T. N. Bhat. 2005. Application of InChI to curate, index and query 3-D structures, *Proteins Struct. Func. Bioinf.*, 60:1–4.

M. D. Prasanna, J. Vondrasek, A. Wlodawer, H. Rodriguez, and T. N. Bhat. 2006. Chemical Compound Navigator: A Web-Based Chem-BLAST, chemical taxonomy-based search engine for browsing compounds, *Proteins Struct. Func. Bioinf.*, 63:907–917.

A. Ruttenberg, T. Clark, W. Bug, M. Samwald, O. Bodenreider, H. Chen, D. Doherty, K. Forsberg, Y. Gao, V. Kashyap, J. Kinoshita, J. Luciano, M. S. Marshall, C. Ogbuji, J. Rees, S. Stephens, G. T. Wong, E. Wu, D. Zaccagnini, T. Hongsermeier, E. Neumann, I. Herman, and K. H. Cheung. 2007. Advancing translational research with the Semantic Web, *BMC Bioinformatics*, 8(Suppl 3):S2.

H. S. Rzepa and P. Murray-Rust. 2001. A new publishing paradigm: STM articles as part of the Semantic Web, *Learned Publishing*, 14:177.

B. Smith, M. Ashburner, C. Rosse, J. Bard, W. Bug, W. Ceusters, L. J. Goldberg, K. Eilbeck, A. Ireland, C. J. Mungall, The OBI Consortium, N. Leontis, P. Rocca-Serra, A. Ruttenberg, S.-A. Sansone, R. H. Scheuermann, N. Shah, P. L. Whetzel, and S. Lewis. 2007. The OBO Foundry: coordinated evolution of ontologies to support biomedical data integration, *Nat. Biotechnol.*, 25:1251–1255.

B. Smith, W. Ceusters, B. Klagges, J. Kohler, A. Kumar, J. Lomax, C. Mungall, F. Neuhaus, A. L. Rector, C. Rosse. 2005. Relations in biomedical ontologies, *Genome Biol.*, 6:R46.

L. Soldatova, A. Clare, A. Sparkes, and R. D. King. 2006. An ontology for a robot scientist, *Bioinformatics*, 22:464–471.

L. N. Soldatova and R. D. King. 2006. An ontology of scientific experiments, *J. R. Soc. Interface*, 3:795–803.

S. Teufel and M. Moens. 2002. Summarising scientific articles: experiments with relevance and rhetorical status, *Comput. Linguist.*, 28(4):409–446.

Part IV

Involving the Researchers
and Closing the Loop

9 The Future of Searching for Chemical Information

David J. Wild and Roger Beckman

CONTENTS

INTRODUCTION

In the previous chapters, the problem of information overload was discussed, and a variety of technical methods of extracting, linking, and mining chemical information were introduced. In this chapter, we discuss the mining of chemical information from the researcher's point of view, particularly how academic and industry researchers currently find the information they need, difficulties and unmet opportunities in mining information, and, finally, some examples of how new technologies may help researchers manage the overload of information in the future.

In many ways, the field of chemistry is extremely fortunate in having had powerful, specialized searching tools like SciFinder Scholar, STN Express, and MDL Crossfire (for Beilstein and Gmelin) available for many years. These tools permit a high degree of flexibility in searching; for example, one can find journal articles that refer to compounds containing particular substructures, search for physical properties of particular compounds, and so on. However, these are bounded tools:

They work on specific, curated databases. We are now entering an age in which one expects to be able to obtain relevant information in an instant and with minimal searching effort, and for this searching to be inclusive of all kinds of information. Ease of access and inclusiveness are often considered before quality and curation. Thus, we are in a situation where we have excellent, specialized tools that only work with limited data sources and an overwhelming mass of data sources on the Web that only have generic searching capabilities. We believe the next leap forward will come when quality, curation, and specialized searching are combined with the wealth of information that is becoming available on the Web and in other sources.

SURVEYING THE USE OF CURRENT TOOLS

To properly understand the likely future trends in searching, one must first consider the current situation. Particularly, how do researchers of chemical information go about finding the information they need? What works well? What does not work well? How does their searching differ from searching of other kinds of scientific information or universal kinds of searching such as that provided by Google?

To answer some of these questions, we have drawn on our own experience (one of the authors, Wild, is a researcher in cheminformatics; the other, Beckman, is head of the chemistry and the life science libraries at Indiana University), as well as conducting informal interviews with chemists in academia who have a wide range of backgrounds. Below we report on interviews with four academic chemists covering their current practices in finding the information they need, what they think works well at the moment, and, of particular interest for this chapter, what they would like to be able to do but currently cannot. We take the liberty of reporting our notes from the interviews in a rather verbatim fashion, in the hope that in doing so we will preserve the small insights that come from the details as well as the more general points.

SUBJECT 1

Our first subject is an academic chemist with prior pharmaceutical industry experience. His scientific activities are focused on the relationship of protein structure and function, with a particular interest in novel methods of drug delivery. He works in a senior position with biochemists, bioorganic chemists, chemical biologists, and proteomics researchers. His characterization of the main difference between the pharmaceutical industry and academia is mostly in the patent area: Academics tend not to be concerned with searching patents, although a lot of information is buried in these documents. Particularly, one can look at the claims, and the order of the claims often gives a hierarchy of the importance of compounds. Currently, he searches the standard databases for published literature for authors or titles for his research. In teaching he has found that Google is useful, for example, to find particular reagents in a less time-consuming fashion than fingering through textbooks. He maintains his own database of about 2,000 articles in EndNote. He considers that this is an "old fashioned" way to do this because it is redundant; that is, he could find the full text of these documents on the Web. He has the hard copy of many of these articles, and this is useful when articles in more "obscure" journals are not available electronically at Indiana University or are not yet digitized.

Due to his senior position, he has colleagues who run citation searches of several seminal publications on a regular basis. They alert him if he should take note of any. Generally, there is a lot of use of direct communication between researchers to share information on relevant articles, particularly using e-mail. This subject subscribes to 12 to 16 print journals; interestingly, this is primarily to enable him to tear out good articles to read while traveling. He posts those he thinks are of general interest to his group on a bulletin board. He keeps articles he has torn out in a folder for a couple of years in case he needs an idea for a current topic. His goal is to get students to read the primary literature. He considers that current students have depth of knowledge but often lack breadth. Strikingly, his greatest fear is not being aware of what he isn't able to do or what he doesn't know. He would like to be able to take structural information and organize it with conformational descriptors and to organize literature by study as well as structure: It might be possible to see the links of study A to study B, but the links from study A to C are not apparent.

Regarding his concerns, he asks whether the results are getting out there any quicker in the current electronic but still journal-centric environment. He feels that the situation is still based on the past and is conservative—a cultural problem and not a technical one. He feels that if blogs could be used to post findings and bypass the traditional editorial process, it would allow results to be disseminated quicker and would allow "controversial" ideas to have a better chance of being brought forward. There is the question of how to validate research in such an environment and how it would be retrieved. We asked him about the library's role in curating datasets because this is a hot topic among library administrators. He says there is a difference in academia, where people work relatively independently, and industry, where a large number work together on a one project. Industry is forced to share more because of the nature of the project. He sees biology and chemistry speaking two different languages and does not see an easy way of bridging the gap.

SUBJECT 2

The focus of our second subject's research is the use of organic synthesis to study problems of biological and medicinal interest. Almost all of the time, he finds the information he needs by searching SciFinder Scholar for authors, subjects, or compounds. He does this kind of searching once or twice a week. He does not currently use Beilstein and thinks he should make use of PubMed too. Maybe 2% of the rest of his information comes from Google and Web of Science (WOS). He is usually looking for information by topic, and to a lesser extent he is looking for information on a compound (just compound, as reagent, prep, or if compound is known). He searches by author the least. The information is used for classroom instruction, writing papers, and preparing grants. He has some paper references in his office, for example, about six volumes of Houben-Weyl that deal with his research.

This subject does not consider that there is information available that he cannot access. Partly, this is because he relies on and trusts the library searchers at the university (he has previously been at institutions where expert searchers are not available and it is hard to find out who to contact for help).

Subject 3

Our third researcher covers the area of molecular nanoscience and nanotechnology. WOS is his primary, all-purpose way of finding chemical information. He can do this quickly and get an overview of a field. Using WOS, he can create a "paper trail" and then mine the papers for the information he needs. He uses SciFinder Scholar only for structure searching. He sees its strengths as finding reactions, locating commercial sources for chemicals, and finding how to make compounds. He sometimes uses the Cambridge Structural Database (CSD) and plans on using the Inorganic Crystal Structure Database (ICSD). Secondarily, he uses textbooks to find information. He does not use Google unless he is looking for something about industry. This subject reads the primary literature that supplies most of the information that he needs (such as property data).

His main current complaints are that he cannot look for structures in WOS. He would like to have a plug-in that would retrieve all the full text of articles that he has located in a WOS search. Currently it takes lots of effort and clicks to go from the citations discovered in a search to pull in all the articles, and when they are downloaded, they have cryptic filenames. He feels it would be useful to have available a history of his search activities along with the search terms that would be automatically stored as the day went along. The ability to call up a previous search history in the rare instance that he would want to find again something he had a hard time locating earlier would be useful.

Techniques get in the way. He asks whether DOIs (digital object identifiers) can be downloaded into EndNote along with article citations without going through lots of steps. Having an online index to all textbooks would be useful whether he has access to the full text or not. He is often looking for something and needs to know in what book and on what page that term occurs. Such an index would need a good filtering ability and display of results.

Another way he approaches the literature is by research program. He finds this to be a good way to organize information. He knows the players in the field or uses WOS if he does not to find out who the major players or research labs are. For him, an ideal tool would be one that would give him the ability to drop PDFs on an interactive map of the world that showed the research labs. When he wanted information on a topic, he would be able to mouse over the map and get information on the various labs' research (probably the material he stored and additional information too). He thinks lots of researchers view the information world in this manner. He does not like the way SciFinder Scholar displays the bibliographic results from a keyword search, and that is one reason he uses WOS.

Subject 4

Our final subject is a synthetic organic chemist who focuses on the preparation of complex substances, for example, natural products of pharmacological significance.

He does not search for information relevant to his field himself but relies on the students and personnel in his lab. They use the standard databases such as SciFinder Scholar and MDL CrossFire. They look by topic or by structure and search based on either an exact structure or a substructure. For substructures, it could be a family of compounds that are of interest; for example, they could be related in some way. If they are looking by starting material, it could give clues on related compounds of interest. If searching for a product (and many of these are unknown), they might look for clues on inherent reactivity or stability that apply to what they are trying to synthesize. This type of searching is very important to them. Other times they are looking for how variations are affected by temperature and other reaction variables. They also do reaction searching and name reaction searches. Searching by reagents is sometimes difficult because it is hard to narrow well since there are often lots of answers.

Usually they get too few hits; this is the nature of the discoveries they are working on. They are at the cutting edge, and the exact thing that they are working on is not in the literature. So they need to look for tangential information or get hints from related work. In searching for information, his past knowledge is often useful. He will often remember a researcher working in a similar area from many years ago. Then his students can look for it, and a citation search will identify newer articles that cite this work. Sometimes his favored technique is to broaden the search and then winnow out the useful information.

He cannot think of anything that they want to do but cannot. He does see a change in attitude of many students today from his education in the 1960s or even in the 1980s. Modern-day students tend to be overly concerned that research they are doing might not be truly unique, to the extent that some may feel embarrassed or threatened on finding that aspects of their work has already been considered in the literature. He recalls in his own work finding a footnote from Faraday that related to a project of his, and he was pleased that it gave him a connection with the giants of the past; the words might be different, and it was exciting to try to understand how the earlier scientists thought about these things. He tries hard to get students to move beyond just looking for "recipes" as the answer and to consider the broader picture. The ease of publishing today may contribute to this attitude.

He also commented on the first subject's information-seeking situation. He said that in that case there are a large number of peptides, and Subject 1 is trying to find those few gold nuggets with drug potential in a big resource pool. Informatics could be a help in that search. Subject 4's situation is different. He is essentially like an architect, needing newly designed materials and techniques to achieve his vision, but on a molecular scale.

Points for Consideration

As we said before, we have reproduced our interview notes in a verbatim fashion in the hope that some useful insights can be found in the details. We also think that some general trends emerge from these interviews:

- In the relatively "standard" areas of chemistry the traditional information sources serve the informational needs of the researchers well. They seem to choose a favorite source and then become proficient at extracting the information they need. Along with their favorite source or sources, they have developed informal information networks or communities and understand very well the flow of scientific information, so they know what they are looking for and have a good sense of how to go about finding it.

- One shared worry that came out in our interviews is that students do not understand the structure of scientific information enough and do not appreciate its breadth. Many students have a limited and sometimes selfish view of the communication function of scientific information.

- The roadblocks to the future are more technical or have to do with options for access, rather than deriving from a basic lack of information. However, this seems to change as one gets away from the "standard" chemistry areas and into the interface of biology and chemistry. There the perceived roadblocks are the inadequacies of the database interfaces; until the information becomes more easily shared, major barriers will continue to exist. Commercial database creators want to protect their investments, so they restrict the number of records that can be downloaded, making it difficult to manipulate and mine hundreds of thousands of hits at once.

These interviews hint at one of our contentions, that this is in many ways the best of times and the worst of times for finding information. Curated databases have proved extremely useful, particularly when used in partnership with specialized librarians, and can be supplemented with Web searching. So much information is available electronically, but the ease of finding "something" often leads to much information being overlooked, particularly for students who are poorly equipped for and inexperienced in finding the information they need. The trends in the academic world are increasingly toward end-user searching, which is certainly convenient and inexpensive, but bypasses the experience of librarians and the information depth and quality of specialized tools. Until the advent of SciFinder Scholar, structure searching on the Chemical Abstracts database was done using complex keyboard commands in the CAS Registry File or software such as STN Express that allowed the searcher to upload the search created offline. At that time, the cost of the search was determined by the time it took to complete and then by the number and types of answers. Users often requested help from an expert librarian. Those days are long gone: The vast majority of searching in academia is now done by chemists on their PCs. Expert searchers are often valued in industry because the result of missing something could be very costly; in the academic world the repercussions are not as extreme (maybe lost time or a lower grade in contrast to law suits or lost profits).

THE ADVENT OF ELECTRONIC ARTICLES AND BOOKS: THE LIBRARIAN'S PERSPECTIVE

Here are some observations from the librarian's perspective at Indiana University. A relatively recent change is the number of journals that are available in electronic form: A chemist can have access to almost all the journal literature from the late

1990s to the present. Most large publishers are selling the back-files of their journals back to volume 1 for one-time prices. The journals of many publishers are available electronically back to issue 1. Users have an alternative search option and can search all a publisher's journals for keywords from their web pages. Having all of these journals online leads to another powerful innovation: the ability to link from a bibliographic citation found in a database to the full-text article. ChemPort is an example, as is SFX (Livingston et al. 2006). Much of the work is behind the scenes in the technical services areas in libraries, along with pointing out errors from the staff working in public areas to keep the knowledge base of subscriptions up to date. The journal literature is available electronically in most libraries, and the trend is toward electronic only. The ability to browse a stack of journals to get ideas for a grant proposal may soon be a thing of the past in all but the best-funded libraries.

There is also a slower but growing trend to move books to online only (Christianson and Aucoin 2005; Just 2007). Part of this is because a printed book is still easier to use for most people (Christianson 2006). There are also issues that impede users from knowing what electronic books are available in the local online catalog as well as poor interfaces for products that aggregate electronic books from many different publishers. Often these aggregators are restricted by publishers and impose a limit of printing or downloading one page at a time instead of presenting a PDF of a whole chapter. Often these products are more of a rental than permanent access as is the case for a printed book. In the future the way books are acquired either in print or electronic version could change. It might be more user-driven instead of collections built through approval plans and individual selection by librarians.

As mentioned above, it is difficult to know exactly what electronic books and journals an institution has access to because it is usually a mixture of "owned" items, "rented" items, and "freely-available" items. The online catalogs provided by institutions are often nonintuitive, particularly in comparison to the highly intuitive Web searching tools available, and this limits users from easily knowing what their institution makes available for them. The capabilities of most online catalogs are primitive in comparison to SciFinder Scholar. Even if one finds a reference in a database to a book chapter, the linking program often does a poor job determining whether a library owns the "book" or series title. The linking programs do a much better job with electronic journals, but our experience leads us to estimate that the success rate is around 95%. The success is very dependent on the quality of the "knowledge base" behind it. There are some new products such as WorldCat Local that include a library's actual holdings along with an extensive collection of full-text articles. These types of products would serve students more than researchers. With users used to Google-like searches that attempt to display the most relevant items first, online catalogs can be frustrating. Title searches can include hits coming from records that contain search terms from the tables of content in addition to the book title. Often the display of records makes no sense to the user or the librarian. Maybe better metadata or allowing users to define fields better would help. One step toward better access may be the concept of functional requirements for bibliographic records (FRBR) (Zhang 2007).

Somewhat related to this is the idea of federated searching. This might be the Holy Grail for users. Put a query for wanted information in a search box and then get all relevant information back in a relevance-ranked list. This has led many libraries to offer federated searching so users can search many different web-based databases at once, although the searching is at a very basic level in comparison to using the native search interface. Many database venders provide the ability to search multiple databases that they provide. Web of Knowledge is one example, as is Scopus. Federated searching is limited in the field of chemistry because so many of the important resources are not web based but often run on software that has to be downloaded to a local workstation. The Indiana University Libraries (http://www.libraries.iub.edu/) are using the federated search model on the libraries' web pages to provide users a resource discovery tool in addition to the standard links to known resources.

One potential drawback with something that is easy is that it does not always deliver the needed information unless the correct term is searched for, and often many results have to be examined to find the answer, if it is available. However, a system that is complicated will frustrate many users. It is still important that users know the structure of information so they can select the appropriate search system or search strategy. The ideal of having everything in one big pot means there have to be adequate filters that will help the user winnow out what is sought. A recent example may help illustrate my point. I (Beckman) was looking for the melting point of a lipid but having no luck in the standard databases. When I consulted a lab that works on lipids, they directed me to a web page of a lipid supplier that they use for information on the lipids they work with. Sometimes something very specific is the best place to go.

So, some trends seem to be selection of a few systems (either by market consolidation or users who rely on a database they are comfortable with) that include multiple databases or a combination of bibliographic and property and chemical data and linking programs that tie primary literature with other forms of data or secondary and tertiary literature.

It is highly probable that the future of the journal will be entirely electronic. Here at the Indiana University Chemistry Library and Life Sciences Library, the journal collections are rapidly heading toward online-only subscriptions. It is rare that we would subscribe to a print-only journal, and they make good targets when we need to cancel titles to balance our serials budgets. Some chemists still find time to come to the library to browse the print journals, although they will find few to browse in 2008. Several have told me that they like to browse through a pile of issues when looking for ideas as they are preparing grant applications. Most researchers receive tables of contents of their favorite journals as they are issued. One researcher reported that he looks at the paper issues of "second tier" titles or titles in journals he does not read on a regular basis. Browsing is still possible in the electronic environment, for example, going to a publisher's website; SciFinder Scholar has an option to "browse my favorite" titles from one's computer or retrieve citations from a journal from an article database that indexes journals cover to cover. At Indiana University the electronic book, at least in a version that is easy to read, print, and download, along with the electronic series, lags behind the electronic journal in moving away from the print version. Although open access will surely loom large in the future, it is unclear how much of the scientific literature will be open access and how efficiently that information can be retrieved if it is deposited in multiple digital depositories.

The recently announced law that all articles resulting from NIH-funded research be accessible from PubMed will likely be a good test of the future of open access (Carroll 2008). A few publications hint at the uneven future ahead (Baker 2006; Borgman 2007; Drake 2006; Hangarter 2005).

WAITING FOR THE GREAT LEAP FORWARD

Coping with the large and increasing amounts and kinds of information available in a meaningful way will necessarily involve a degree of automation of some traditionally human activities. One only has to do a search with Google to reveal that technology has advanced to a point that it is possible to automatically aggregate and search large volumes of diverse kinds of information in a humanly manageable form. The search engine now spans very different kinds of data (online books, journal articles, news items, images, videos, and so on), and when one performs a search, it makes a reasonable attempt to present the results to highlight the items most likely of interest, and makes it very easy to restrict the search to particular domains or items. For example, if one searches for "William Shakespeare," one is presented (at the time of writing) with a top hit that lists online scanned books by the author, then Wikipedia pages describing his life, and then other biographies, study notes, and so on. One can quickly modify the search to only include online books, or if one is really interested in pictures of "the Bard," then one only has to click on the images link to restrict it in this fashion. One can dig deeper to find videos, news items, and so on.

We believe that it is perfectly possible for searching for chemical and related information to be this simple. For example, one might be interested in information on quinazolinediones for a number of reasons: a desire to know the melting point of a particular compound, an interest in medicinal chemistry papers on the biological properties of these compounds, an interest in groups researching these compounds, and so on. At the moment, such a search on Google provides an overload of poorly ordered hits, but there is no reason why the same organization that is applied to more popular searches could not be applied in the chemical and life sciences. However, a number of serious hurdles have to be faced by the scientific community before such a tool would become feasible.

LIBERATION OF INFORMATION IN JOURNAL ARTICLES

As discussed earlier, the future of journals is clearly electronic. However, there are many reasons why journal articles are problematic in comparison to other documents such as web pages and databases. First, most chemistry journals (with the notable exception of *Chemistry Central Journal*, www.journal.chemistrycentral. com) are not open access, and thus the content of articles is restricted by the publishers. Although most universities and large organizations have institutional subscriptions to the popular journals, access usually requires validation on computer IP addresses or the use of private login credentials. Thus, automated access to this information by a computer is difficult. Further, it is unclear whether the terms under which journal articles are made available permit automated processing of the content (for example, creating an index of terms used in the paper); the intent of permission terms does not seem to address these kinds of uses specifically.

Second, even if one can access the content of a journal article page, it is generally not in a format that makes it easy for a computer to process the information contained in it. In particular, the PDF format is most commonly used, in which chemical structures and spectra become images, and data tables become plain text. Maybe a journal article refers to a melting point of a compound that would answer the query of a particular scientist, but how would a computer locate this information in the text? The cause of creating machine-readable versions of journal articles and other documents in the chemical sciences has been championed by Peter Murray-Rust and Henry Rzepa (2004), who have coined the term "datument" to refer to a document that is fully marked up in machine-readable form, so that meta-information about data is fully preserved.

Third, chemistry is a subtle art, and journal articles often contain all kinds of intricate information about compounds, reactions, and their properties in very concise form. It is for these reasons that human curation of journal articles has been so important in the success of tools like SciFinder Scholar. For the information in journal articles to be freed to the world of uncurated searching, major policy changes have to be made in the way journal articles are assembled, formatted, and accessed. One assumes that open-access journals will lead the way in this. It is notable that *Chemistry Central Journal* does plan to include International Chemical Identifier (InChI) format representations of compounds referenced in papers with the articles themselves. Additionally, the Royal Society of Chemistry is experimenting with the provision of articles that are marked up with both chemical structure information and highlighting of ontological terms (Gene Ontology and GoldBook) and is making some of these articles available through Project Prospect (http://www.projectprospect.org).

Metadata

This hurdle involves the ability to assign and retrieve meta-information pertaining to quality, confidence, curation level, and source of information. There are straightforward technical ways of storing metadata (information about data) along with data and the content of documents: In particular, XML and microformats enable markup of web pages and documents by tagging parts of information. For example, to represent a boiling point, one could surround the value with tags <BoilingPoint>100</BoilingPoint>. However, this is not as simple as it looks. How do we standardize the metadata names (BoilingPoint vs. BP vs. Boiling_Point, etc.)? How do we represent units (100 centigrade, Fahrenheit, kelvin)? How do we specify context and parameterization (conditions etc.)? The issue is even more complex if we want to express less tangible properties such as confidence in a result, quality, amount of curation that has been applied, provenance, and so on.

Strong Security Where Necessary

Security is an issue in both academia and the pharmaceutical industry, although it is of greatest concern in industry, where a security lapse can in the worst case cause a highly expensive loss of competitive advantage. In academia, the consequences are less onerous, although most scientists are eager to protect their own intellectual property, particularly in the fragile early stages of research. Thus, for searching systems to be widely used, security has to be addressed. While many security issues are generic, one specific to this community is the transport, storage, and use of chemical structure and derivative information. A researcher at a pharmaceutical company might itch to search public databases for compounds similar to a "hot" new one recently synthesized and found to be biologically active, but if this involves submitting the structure to be (even temporarily) stored as a query on remote Web servers, company policy if not personal concern is likely to prohibit him or her from doing so. Pharmaceutical companies are thus used to closed systems, where most information is generated (or at least held) internally to the company, and searching systems span only servers behind the company firewall.

However, future drug discovery is clearly going to have to make use of information that is public (or at least not proprietary to the company) and is only accessible outside of the company firewall. This may be achieved by either copying all pertinent information inside the firewall, updated on a regular basis, or by developing secure methodologies for searching outside sources without revealing structural information. The former strategy is not as difficult as it seems (disk storage is cheap, and internet bandwith wide enough to enable transfer of large volumes of information daily), although it results in a significant restriction in the kinds of searching that can be performed, as any new search tools or techniques also have to be imported into the company and modified for use on the internal versions of the data. The latter approach may be tackled either by using secure (secure sockets layer) connections with some method of validating deletion of query structures once searches are complete or by encoding the structural information in a way that it cannot be reverse engineered back into a chemical structure; for example, a carefully designed fingerprint might be used as a similarity search query without need to directly provide chemical structure information (although it is possible the structure could be estimated from the top hits from the search).

An Open Lab Culture Where Possible

Unlike the security issue, which is technical, this is more of a cultural hurdle. Large amounts of useful information are generated by (particularly academic) chemistry laboratories but never published (or publication is delayed). This can be for a number of reasons: The information might pertain to negative information not considered useful to the research project; there might be intellectual property concerns; the information might be not considered publishable yet in journals. The problem of publication bias (the tendency to publish only positive results) is widely studied, and any solution goes beyond the scope of chemical information. Intellectual property concerns would likely require information privacy. Publishing key research findings outside of journals may be considered prepublication by the journals. However, we do think there is scope for much more information being made available, particularly on the Web. Although some information is always going to be preferred to be kept private, not all information generated in a project or in a lab needs to be kept private.

Laboratories constantly generate pieces of information that do not make it to journal articles (or if they do, are incidental to the thrust of the paper), yet could prove useful to other researchers. It is possible that the only reason this information is not provided by laboratories is that there is not a convenient venue for disseminating it (and for searching information provided by other groups and labs). For this reason, we find emerging projects such as Open Notebook Science (http://precedings.nature.com/documents/39/version/1) and O2HU (http://www.o2hu.com/) highly interesting. We believe more study of lab culture associated with innovation in the use of the web is required.

If these issues can be addressed, then integrated searching becomes a fairly straightforward technical issue. Wild (2006) introduced a four-layer model of future life science information storage and use that is scalable to large volumes of information and that tackles some of these points. The layers are, as follows, with the main connection point to the chemical researcher being in the fourth layer:

1. **Storage layer:** Comprehensive information storage including semantics and metadata. May be in a single system or multiple systems.
2. **Interface layer:** Common interfaces to stored information. There may be several for different kinds of information.
3. **Aggregation layer:** Software, intelligent agents, and data schemas customized for particular domains, applications, and user.
4. **Interaction layer:** Software for information access and storage by humans, including e-mail, browsing tools, and "push" tools.

The World Wide Web, in particular "Web 2.0" and Web services, are affording significant improvements at layers 1 and 2. It is now possible to store vast amounts of information on web servers (and if one does not have the capacity for this on one's own servers, it can be cheaply purchased through services such as Amazon's S3 that can be quickly accessed over high-speed networks. Standardized databases including Oracle, MySQL, and PostgreSQL (with network access made easy through JDBC and ODBC) and Web service interfaces permit direct access to the data in organized, searchable form (indeed, even specialized structure, substructure, and similarity searching can be easily implemented in these databases with cartridge tools). XML, microformats, and Web service interface descriptions are beginning to enable standardized representations for data and means of access. However, little work has yet been done on layers 3 and 4; in particular, the aggregation layer is often skipped. The aggregation layer is the primary layer that provides scalability: Information is aggregated and filtered so it can be delivered to the human in manageable quantities and forms. Without this layer, data deluge quickly ensues. Many journals already offer e-mail alerts that allow a researcher to receive e-mails if articles are published in journals that meet specific search criteria. However, these alerts require search terms to be strictly defined, and their scope is usually just to one publisher's products. The use of intelligent agents, and intelligent ways of pre-organizing and filtering data, deserves much more attention than it currently receives.

Despite concerns about prepublication of scientific results and plagiarism, an increasing amount of chemical information is being placed on the Web, in chemistry blogs, online documents, and web pages. The Useful Chemistry blog (http://usefulchem.blogspot.com/) encourages posting of chemistry problems and their solutions, under a creative commons license. At the time of writing, it has a fairly small number of contributors (16), but it is widely read. Other blogs such as Chem-Bla-Ics (http://chem-bla-ics.blogspot.com/), Noel O'Blog (http://baoilleach.blogspot.com/) and PeterMR's Blog (http://wwmm.ch.cam.ac.uk/blogs/murrayrust/) are used for reflection on the fields of cheminformatics and chemical information handling. A more ambitious development is the Open Notebook Science project at Drexel (http://drexel-coas-elearning.blogspot.com/2006/09/open-notebook-science.html), which encourages scientists to record their findings in an openly accessible format, such as a Wiki. Interestingly, Wikipedia (http://www.wikipedia.org) has been used to document many common molecules, including properties, therapeutic uses, and machine-readable representations of the two-dimensional structure (including SMILES and InChI).

The Semantic Web extends the current Web and aims to make web content machine-readable, thus allowing conveyance of meaning and integration of data objects in different formats, potentially enabling autonomous agents to carry out various automated tasks by discovering services and information required to complete them. In the Semantic Web community, the Semantic Web Services Interest Group from W3C has taken the lead in developing specifications for this, perhaps the most widely applicable being OWL-S. OWL-S contains a set of generic ontologies to characterize what the Web services do, how they perform the task (including workflow descriptions), and how to access them. It also provides the option to extend the generic ontology with a specific domain ontology that describes entities and services particular to a given domain (in our case cheminformatics).

Once these ontological specifications are created, it is possible to apply reasoning tools to automatically create workflows of services that tackle tasks that require the involvement of multiple services. These are of particular interest, as they offer the possibility of on-the-fly aggregation of services and information in response to a scientist's (potentially complex) query, without the need for workflows to have been predefined. Such reasoning tools already exist, but they require exhaustive search of the Web services space (an NP-complete problem). Techniques and heuristics are being developed in the Semantic Web community to reduce the search spaces and effect efficient searches. We will participate in these efforts while tailoring the searches to cheminformatics.

CONCLUDING REMARKS

We believe much is working well in the world of chemical information. We are fortunate to have highly effective specialized tools such as Scifinder Scholar, a good supply of specialized librarians and chemical information–searching specialists in both academia and industry, and are now able to supplement this with web searching. This is reflected in the general level of satisfaction (at least of academic chemists) with the current situation. However, the explosion of information available in journal articles, public databases, and the Web affords all kinds of opportunities that were not available previously. If some policy and technical barriers are crossed, specialized chemical information searching may in the near future be as straightforward as searching Google is today.

REFERENCES

Monya Baker. 2006. Open-access chemistry databases evolving slowly but not surely, *Nature Reviews Drug Discovery*, 5(Sept.):707–708.

Christine L. Borgman. 2007. *Scholarship in the Digital Age: Information, Infrastructure, and the Internet*. Cambridge, MA, MIT Press.

Michael W. Carrol. 2008. Complying with the NIH Public Access Policy—Copyright Considerations and Options: A SPARC/Science Commons/ARL Joint White Paper. Available at http://www.arl.org/sparc/advocacy/nih/copyright.html (accessed March 7, 2008).

Marilyn Christianson. 2006. Patterns of use of electronic books, *Library Collections, Acquisitions, and Technical Services*, 29(4):351–363, doi:10.1016/j.lcats.2006.03.014.

Marilyn Christianson and Marsha Aucoin. 2005. Electronic or print books: which are used? *Library Collections, Acquisitions, and Technical Services*, 29(1):71–81, doi:10.1016/j.lcats.2005.01.002.

Miriam A. Drake. 2006. A scholarly society faces open access—The American Chemical Society: interview with Bob Bovenschulte, president publishing division, American Chemical Society, *Searcher*, 14(4):8–15.

Roger Hangarter. 2005. Response: letter to the editor, *Wall Street Journal*, June 21. Available at http://www.aspb.org/openaccess/wsjletter.cfm (accessed March 7, 2008).

Peter Just. 2007. Electronic books in the USA – their numbers and development and a comparison to Germany, *Library Hi Tech*, 25(1):157–164, doi: 10.1108/07378830710735939.

Jill Livingston, Deborah Sanford, and Dave Bretthauer. 2006. A comparison of Open URL link resolvers: the results of a University of Connecticut Libraries environmental scan, *Library Collections, Acquisitions, and Technical Services*, 30(3-4):179–201, doi:10.1016/j.lcats.2006.08.001.

Peter Murray-Rust and Henry Rzepa. 2004. The next big thing: from hypermedia to datuments, *Journal of Digital Information*, 5(1):248.

David J. Wild. 2006. Strategies for using information effectively in early-stage drug discovery, in Ekins, S. (Ed.), *Computer Applications in Pharmaceutical Research and Development*. Wiley-Interscience, Hoboken.

Yin Zhang. 2007. Functional requirements for bibliographic records, *Bulletin of the American Society for Information Science*, 33(6):6.

10 Summary and Closing Statements

Debra L. Banville

This book was meant to provide a good starting place for life science researchers who rely on chemical information in their work. Each chapter introduced researchers to key concepts necessary to the understanding of chemical information mining without the need to understand all the technical information presented. For those interested in using chemical mining capabilities, Part II's discussion of chemical semantics provided a good background of the available capabilities, with expert comparatives on the underlying technologies. Part III tied the chemical mining capabilities into the bigger picture of linking chemical information with biological information and the huge impact the Semantic Web is having on how we search for and look at information. Finally, Part IV looked at the current tools available and projected into the future of chemical information searching for both academic and commercial research areas, answering key questions about the unique needs of these different user groups. The experts assembled to write each chapter have presented an array of ideas and concepts that should provoke and challenge the way we look at information now and in the future.

This book will certainly be outdated in some aspects before publication. For some it will provide too much detail in some areas and not enough in others, but it is hoped that this book provides a good initial overview with the necessary references for a diverse set of needs. It is also hoped that this book will contribute to the momentum building toward improved annotation at the point of publication of scientific articles using universally adopted standards, and new business models that can improve the accessibility of research information while the providers of this information can cover their costs in ways that are mutually beneficial to all. The issue of standards is not new to other industries such as in the aeronautics industry to promote improved air safety (Kohn et al. 2000). A similar argument can be made about the safety of pharmaceuticals (Banville 2008). Certainly, literature-based discovery or text mining for the life sciences is here; technologies exist today that offer an improved ability to manage both chemical and biological information and offer researchers a road map to novel discovery.

Finally, this book will end with the words of Donald R. Swanson, professor emeritus at the University of Chicago:

> More than 40 years ago the fragmentation of scientific knowledge was a problem actively discussed but without much visible progress toward a solution; perhaps people then had the consummate wisdom to know that no problem is so big that you can't

run away from it (from Swanson's 2001 ASIST Award of Merit acceptance speech [Swanson 2001]).

REFERENCES

Banville, D.L. 2008. Mining chemical structural information from the literature, in K. V. Balakin (Ed.), *Pharmaceutical Data Mining: Approaches and Applications for Drug Discovery.* Wiley Inc., Chap. 20.

Kohn, L.T., Corrigan, J.M., and Donaldson, M.S. (Eds.). 2000. *To Err Is Human: Building a Safer Health System.* Washington, DC: National Academic Press.

Swanson, D.R. 2001. On the Fragmentation of Knowledge, the Connection Explosion, and Assembling Other People's Ideas. ASIST Award of Merit Acceptance Speech. *Bulletin of the American Society for Information Science and Technology* 27(3). Available at http://www.asis.org/Bulletin/Mar-01/swanson.html.

Index

Milton Keynes UK
Ingram Content Group UK Ltd.
UKHW040057071024
449327UK00019B/630

9 780367 386207